The British
Electricity Experiment

The British Electricity Experiment

Privatization: the Record, the Issues, the Lessons

Edited by John Surrey

EARTHSCAN

Earthscan Publications Ltd, London

First published in the UK 1996 by
Earthscan Publications Limited

A catalogue record for this book is available from the British Library

ISBN: 1 85383 370 3 (Paperback) 1 85383 371 1 (Hardback)

Typesetting and page design by PCS Mapping & DTP, Newcastle upon Tyne

Printed and bound by Biddles Ltd, Guildford and Kings Lynn

Cover design by Dan Mercer

For a full list of publications please contact:
Earthscan Publications Limited
120 Pentonville Road
London N1 9JN
Tel. (0171) 278 0433
Fax: (0171) 278 1142
E-mail: earthinfo@earthscan.co.uk

Earthscan is an editorially independent subsidiary of Kogan Page Limited and publishes in association with the WWF-UK and the International Institute for Environment and Development.

Contents

III The Issues

IV A Model to Follow?

List of Tables

List of Abbreviations

ACORD	Advisory Council on Research and Development
AEB	Area Electricity Board
AES	Associated Energy Services
AGR	advanced gas-cooled reactor
BATNEEC	Best available technology not entailing excessive costs
BC(C)	British Coal Corporation (formerly the National Coal Board – the NCB)
BEA	British Electricity Authority
BG	British Gas plc
BMFT	Ministry of Research and Technology (Germany)
BNFL	British Nuclear Fuels Limited
BPO	Best Practicable Option
BST	bulk supply tariff
BWR	boiling water reactor
CCA	current cost accounting
CCGT	combined-cycle gas turbine
CEGB	Central Electricity Generating Board
CHP	combined heat and power
CFDs	contracts for differences
CO_2	carbon dioxide
DGES	Director General of Electricity Supply (the regulator)
DNC	declared net capacity
DEn	Department of Energy
DTI	Department of Trade and Industry
EC	European Commission
ECC	Electricity Consumers' Council
EDF	Electricité de France
EFL	external financing limits
ENOR	existing nuclear operating regime
EPRI	Electric Power Research Institute (USA)

ESI	electricity supply industry
EU	European Union
FFL	fossil fuel levy
FGD	flue gas desulphurization
GJ	gigajoule
GW	gigawatt (1 GW = 1000 MW = 1 million kW) (a unit of capacity)
HDR	hot dry rocks
HMIP	Her Majesty's Inspectorate of Pollution
HSE	Health and Safety Executive
HTR	high temperature reactor
IPPs	independent power producers
kWh	kilowatt hours (a measure of electrical output)
LCPD	Large Combustion Plant Directive
LOLP	loss of load probability
LRMC	long-run marginal costs
MANWEB	Merseyside and North Wales Electricity Board
MMC	Monopolies and Mergers Commission
mt	million tonnes
mtce	million tonnes of coal equivalent
MW	megawatt (1 MW = 1000 kW) (a unit of capacity)
NAO	National Audit Office
NE	Nuclear Electric
NEC	net effective cost
NFPA	Non-Fossil Purchasing Agency
NFFO	Non-Fossil Fuel Obligation
NGC	National Grid Company
NIE	Northern Ireland Electricity
NORWEB	North Western Electricity Board
NOVEM	Netherlands Agency for Energy and the Environment
NSHEB	North of Scotland Hydro-Electric Board
NUM	National Union of Mineworkers
OECD	Organization for Economic Co-operation and Development
OFFER	Office of Electricity Regulation (the staff which supports the DGES)
OFGAS	Office of Gas Supply
OFT	Office of Fair Trading
OFTEL	Office of Telecommunications

OPEC	Organization of Petroleum Exporting Countries
p/kWh	pence per kilowatt hour
PSBR	public sector borrowing requirement
PPP	pool purchasing price
PPSs	purchasing power standards
PWR	pressurized water reactor
pv	photovoltaic
R&D	research and development
R,D&D	research development and demonstration
REC	regional electricity company (responsible for distribution and supply)
RPI	Retail Price Index
SGHWR	steam generating heavy water reactor
SHE	Scottish Hydro-Electric
SN	Scottish Nuclear
SO_2	sulphur dioxide
SP	Scottish Power
SSEB	South of Scotland Electricity Board
SWALEC	South Wales Electricity Company
TISC	Trade and Industry Select Committee (of the House of Commons)
UK	United Kingdom
UKAEA	UK Atomic Energy Authority
VAT	Value Added Tax
VOLL	value of lost load
WEG	Wind Energy Group

The Authors

All the authors work in the Energy Programme of the Science Policy Research Unit (SPRU), which is at Sussex University. SPRU was established in 1966 to contribute, through research, to public understanding of science and technology issues. Accordingly, its research is problem-oriented and therefore inter-disciplinary. The Energy Programme was launched in 1969 and its aim is to inform public debate and understanding on leading national and international energy and environmental issues and their implications for public policies. The Programme has 'centre' support from the Economic and Social Research Council, sponsorship from a long-standing 'club' of industrial companies, and contract funding from the European Commission and other public agencies.

Professor John Chesshire BA (Durham), MA(Sussex). Previous experience with TUC Economics Department. With SPRU Energy Programme since 1974 and Head of Programme since 1986. Member of Electricity Consumers' Council (1977–86). Specialist Adviser to numerous House of Commons Select Committee Inquiries on energy matters (Energy Committee, 1979–92; Welsh Affairs, 1990–91; Employment, 1992–93; Environment, 1993–94; Trade and Industry, 1995; and Northern Ireland Affairs, 1995). Current research interests: industrial energy demand, energy technology policy, and effects of electricity privatization.

Professor John Surrey BSc (Econ) (London). Previous experience with the Central Electricity Generating Board and as government Economic Adviser. Head of SPRU Energy Programme, 1969–86. Specialist Adviser to numerous House of Commons Select Committee Inquiries on energy matters (Science and Technology, 1974–76; Energy 1979–92; Employment, 1992–93; and Northern Ireland Affairs, 1995). Current research interests: market structure, regulation and competition in the electricity and gas industries in the UK and the European Union, UK energy policy (particularly for coal and nuclear power), and nuclear power decommissioning.

Gordon MacKerron BA (Cantab), MA (Sussex). Senior Fellow. Previous experience with governments of Malawi and Nigeria, Glasgow College of Technology, Institute of Development Studies, SPRU (1974–76), and Griffith University, Australia. With SPRU Energy Programme since 1978. Adviser to Monopolies and Mergers Commission Inquiry on the CEGB (1980–81), Adviser to OFFER on nuclear power issues (1994), and Specialist Adviser to the House of Commons Trade and Industry Select Committee on the effects of electricity privatization, 1992–93. Current research interests: economic effects of electricity privatization and the costs of nuclear power.

Steve Thomas BSc (Bristol). Senior Fellow. Previous experience in industry and the Operational Research Department at Sussex University. With SPRU Energy Programme since 1979. Adviser to National Audit Office and consultant to the UK Radioactive Waste Management Advisory Committee on nuclear power plant decommissioning (1993–94). Consultant to the International Atomic Energy Agency (1993–94). Current research interests: restructuring of electricity supply industries in Europe; nuclear power policy in Eastern Europe and the former Soviet Union; the structure and competitiveness of the international power plant industry; and the determinants of nuclear power plant performance and economics.

Catherine Mitchell BA (London), MA (Boston, USA), DPhil (Sussex). Fellow. Previous experience: journalist on oil industry and environmental matters. With SPRU Energy Programme since 1993. Specialist Adviser to House of Commons Welsh Affairs Committee Inquiry on Wind Energy, 1994. Current research interests: the economics, financing and diffusion of renewables and other new energy technologies.

Mike Parker MA (Oxon). Honorary Fellow. With SPRU Energy Programme since 1991. Member of Government's Energy Advisory Panel, and Associate Fellow of the Energy and Environment Programme of Chatham House. For many years previously Director of Economics with British Coal Corporation. Current research interests: the coal industry in Britain and the European Union; the effects of privatization, regulation and competition on the British gas and electricity markets; and UK energy policy.

Isabel Boira-Segarra BA (Open University), MSc (Sussex). DPhil Research Student. Previous experience as a mechanical engineer with the Lucas Group involved in developing new products. Currently researching environmental management within the economic context of England,

Wales and Spain. Commissioned by *The Financial Times* to write a Management Report on The Spanish ESI in 1996.

Jim Watson MEng (Imperial College) in Electrical and Electronic Engineering. DPhil Research Student. Previous experience in the motor industry as an engineering apprentice with Rover Group, and in wind energy technology. Currently researching the success of the Combined Cycle Gas Turbine for a doctoral thesis. While at SPRU, has also carried out work on the privatized electricity industry and on directions and levels of R&D expenditures by governments of International Energy Agency member countries.

Foreword

The electricity supply industry is some 120–150 years old, Faraday having discovered the connection between magnetism and electricity in 1831, though some decades were to elapse before the exploitation of his discovery. The industry's history has been punctuated by acts of legislation, the first, promoted by Joseph Chamberlain, having been in 1881. As he himself put it:

> The supply of gas and of water, electric lighting and the establishment of tramways must be confined to very few contractors. They involve interference with the streets, and with the rights and privileges of individuals. They cannot, therefore, be thrown open to free competition, but must be committed, under stringent conditions and regulations, to the fewest hands.[1]

Accordingly the Electric Lighting Act of 1882 allowed local authorities to break up streets for the laying of cables or to give their consent to private companies to do so, provided for maximum prices and for the purchase by local authorities of companies under the Act at written-down value after 21 years.

As the technology of the industry progressed, the producing unit with lowest costs outgrew the existing units, whether municipal or private, thus rendering them uncommercial. By the end of the First World War there were some 600 undertakings, all with different voltages, the differences making inter-connection difficult. Clearly some consolidation of the industry into larger connected units was required, but the units already in being would not voluntarily relinquish their position. Finally, in 1926 a Conservative Government under Stanley Baldwin passed an Act establishing the Central Electricity Board as the owner of a national grid, or

---

1 Quoted in Leslie Hannah, Electricity before Nationalisation, Macmillan, 1979, p23.

system of cables and power stations, buying power from selected stations, the operation of which it therefore effectively controlled, and selling it wholesale; it also had powers to plan new stations. The first chairman of the Board was Sir Andrew Duncan, for whom I was later to work in the early 1950s when he was independent Chairman of the British Iron and Steel Federation.

The solution of the Board was successful in dealing with the problem of the transmission of electric power from region to region, but it still left untouched the problem of the diversity of supply within regions, a problem which voluntary effort likewise failed to overcome. It was this failure that led ultimately to compulsory ownership by the state under the nationalization statute of 1948, a statute supported by much 'middle of the road' opinion, including Sir Andrew Duncan, although he was opposed to nationalization in general. Under the statute there was established a central authority to operate the generating stations, and to transmit power to semi-autonomous regional distribution boards. The first chairman of the British Electricity Authority was Lord Citrine, a former general secretary of the Trades Union Congress (TUC).

Such was the system which I inherited in the mid-1950s when I was Minister of Fuel and Power. I was to appoint members of the boards and to approve or disapprove of investment programmes. I had no formal power over prices, but I had a power to issue directions 'in the public interest', a power which I never exercised nor have known others to exercise. When therefore Harold Macmillan, as Chancellor of the Exchequer, called for restraint in prices and Lord Citrine, in response, froze the charges of all electricity boards, I was powerless to do anything about it, although I disapproved of Citrine's action.

As Minister for Fuel and Power I was the recipient of the first report ever into the efficiency of a nationalized undertaking – the so-called Herbert report, named after Sir Edwin Herbert (later Lord Tangley), who had chaired the committee of inquiry. I agreed with the Herbert report that the nationalization statute was ambiguous. As I have stated, the area distribution boards were semi-autonomous. What did semi-autonomy mean? Either the distribution boards were fully autonomous or not. Accordingly I drafted a legislative scheme separating generation and transmission, on the one hand, from distribution, on the other, and requiring each board to pay its way. Under my scheme all boards were to confer with each other in a council chaired, like the Gas Council, by an independent chairman, although I was conscious of the fact that the generating and transmitting board would have much the greater weight. It was left to my successor, Lord Mills, to give legislative effect to the scheme.

This was the last major piece of legislation affecting the industry before the denationalization or (to use the new word) privatization statute of 1990–91. This statute was novel in two senses: first, it placed the entire industry (with the exception of the nuclear generating stations) for the first time in private hands, for, as I have shown, much of the industry had been municipally owned; second, it separated transmission from generation, whether for good or ill it is too early to say.

Certain consequences are, however, already clear. Privatization has unintentionally dealt a heavy, probably mortal, blow to the generation of electricity by nuclear fission or fusion, private finance not being able to bear the magnitude of the costs and risks. Secondly, it has inspired a 'dash for gas', the area distribution boards being wary of undue reliance on the separated generators. On both counts privatization has increased the reliance on fossil fuels, with untoward implications for the environment and for global warming. These consequences make me doubt whether the saga of legislation which I have tried briefly to summarize is now finally closed. As I have shown, this is a strategically important industry in which the Government is inevitably involved and there is therefore a complex interaction between politics and economics in the policies adopted for this industry. To those who wish to understand these interactions more clearly, I commend this book.

March 1996

The Rt Hon Aubrey Jones
Minister of Fuel and Power, December 1955 – January 1957
Chairman, National Board for Prices and Incomes, 1965–70
President, Oxford Energy Policy Club, 1976–88

Acknowledgements

This book takes a hard look at British electricity privatization – widely known as the 'British Electricity Experiment' – in order to assess the facts and the full range of issues. Even though all the authors have done considerable previous research in the area, working together on the book proved to be a great mind-clearing exercise. Perhaps inevitably, it generated as many questions as answers.

The book could not have been completed without the considerable cooperation and tenacity of each of the authors, nor without first-class support from our secretaries, Barbara Graham-Carter and Eunice Surtees-Hornby, to whom we express our sincere thanks.

We are particularly grateful to individuals in different organizations for their help and advice on a range of issues and their comments on draft chapters, and to both the Economic and Social Research Council and our 'club' of energy industry sponsors without whose financial support this book (and our research over many years) would not have been possible.

Our final acknowledgement is to colleagues in our host institution, the Science Policy Research Unit, for having provided over many years a very suitable atmosphere in which problem-oriented policy analysis can be carried out, including the issues behind the 'British Electricity Experiment'.

Part I

FROM PUBLIC TO PRIVATE OWNERSHIP

1

Introduction

John Surrey[*]

Objectives and Rationale

The biggest and most radical project in the extensive UK privatization programme took place in 1990–91 when the electricity supply industry (ESI) in England and Wales passed into the private sector. This was the only major public utility privatization which involved significant restructuring in order to promote competition in generation and in retail electricity supply, and to separate transmission from generation. At the same time, a new system of regulation was introduced for the remaining monopoly parts of the industry. The decision against simply transferring the old highly integrated structures from the public sector into the private sector was taken in response to criticisms levelled at the privatization of the telecommunications and gas industries in the mid-1980s which created structurally unreformed private sector monopolies.

The politicians directing the ESI privatization project in 1987–89 set themselves a very tight timetable to convert the whole of the ESI from public to private ownership and to create 17 new companies in England and Wales alone. This timetable involved major risks, as illustrated by the withdrawal of nuclear power from the sale in 1989. But the same politicians were more risk-averse in setting the timetable for the transition to full competition in retail electricity supply, and allowed eight years for this.

* The author is grateful to his colleagues, especially Mike Parker and Gordon MacKerron, for their inputs to this chapter.

In political terms, the final stage was put off until two general elections had passed, when a new generation of ministers would be in post. This most complex and testing stage of introducing supply competition in the domestic or residential electricity market is planned to take place in April 1998. This emphasizes the unfinished nature of the agenda and the large uncertainties involved for a considerable period into the future.

The impending abolition of the industry's territorial franchises, which the final stage requires, is a step into the unknown. Since its beginnings over 100 years ago, the ESI in all countries has been organized mainly on the basis of utility companies vertically integrated or linked by common state ownership, governed by an obligation to supply all consumers in their franchise area (sometimes a given locality, sometimes a region, and sometimes a whole country), and to do so without 'undue preference' or price discrimination. This was combined with some form of regulation to curb their monopoly power. In return, the utility companies were able to pass through capital, fuel and other costs to final consumers and in particular recover their capital expenditures in full (including a rate of return acceptable to the owners).

Few industrial investments are more costly to build than major power stations and power grids. The main benefits of this basic model (there were many detailed variations in practice) were security of supply for the individual consumer and for consumers collectively, combined with an incentive to invest in new plant and equipment, since vertical integration and cost pass through avoided the risk of stranded assets (power stations with insufficient demand to allow their capital costs to be recovered).

Even at the time of writing (early 1996), with only two years remaining and, as yet, no firm plans for market tests for electricity (unlike gas where market tests are due to start in the south west region in April 1996), there is much uncertainty about: whether the two industries will be ready in time; whether the big computerized trading and settlements systems will work reliably; what proportion of consumers will exercise their new-found choice; how far consumers will be protected and whether both service standards and important social obligations to the elderly, the disabled and the poor will be maintained; and how many new suppliers will enter the new electricity supply market to compete with however many of the original 12 regional electricity companies (RECs) may remain after the wave of takeover bids and mergers which began in 1995. There are many such imponderables, not least being the likely ratio of 'winners' to 'losers' among domestic consumers and whether overall supply security under market competition can be maintained at the same high level that consumers have been used to since the late 1940s.

As many elements have yet to be played out, any assessment is necessarily tentative in certain respects. But an assessment at this time can be based on the accumulated record and a far better understanding of the strengths and weaknesses of the new regime than the somewhat polarized claims and counter-claims made in the heat of the British privatization debate in the period 1988–91. Such an assessment, before the final, most radical stage occurs, seems to us highly desirable. We therefore felt that it would be well worthwhile to undertake a wide-ranging examination of the record of the privatized ESI on the basis of the almost six years of experience which has built up so far, and the prospects for the future as far as we can judge them.

There are also three more specific reasons for this book. Firstly, the privatization programme under the long, uninterrupted period of Conservative government which began in 1979 when Margaret Thatcher became Prime Minister and is still continuing under John Major, is by any test a momentous stage in the economic history of the UK. In terms of capital employed, administrative complexity, the number of redundancies it has caused, its effects on wider policy objectives, and the profitability of the various parts of the industry, the modern privatization programme has been a vast task which has occupied the Government over most of its 17 continuous years in office since 1979. It has indeed been the very long period of uninterrupted Conservative government which has permitted the immense scope of the programme – and it has had major and varied impacts. With a book value of assets of at least £42 billion, on a current cost basis, the ESI was the jewel of the whole privatization programme.

Secondly, electricity privatization – more than any of the other privatizations – has been borne along on the intellectual and ideological trajectory of the New Right to the point at which privatization and competition appear to have achieved the near-total eclipse of the case for retaining public ownership. There is even less discussion of re-introducing public ownership of the utility industries on the basis of monopoly provision under the old Morrisonian style of public corporations, statutorily independent of the state but obliged to carry the interest of the state in various respects, dependent on the state for investment finance and at the same time accountable to Parliament for both strategic decisions and often for aspects of their detailed operations. But like any new conventional wisdom the new intellectual framework, which is described in the next section, must be examined critically. This is even more necessary given the ideological slants on interpretations of the record so far and of the unresolved issues which continue to fill many newspaper columns in the UK and more widely. The debate about privatization and competition, especially with regard to electricity, is in fact being carried on internationally as well as nationally, and it is an important aspect of the wider question of what is needed to reinvigorate capitalism in the West and replace communism in the East.

A third motive is to inform people in as balanced a way as possible of the British electricity privatization experience so that they can put into perspective the idealized versions one hears sometimes from international consultants and even World Bank staffers. Many other countries are either treading the path of electricity privatization already or they are contemplating doing so for various reasons. They should not accept the British model uncritically. Both differences in the objectives of privatization and the uniquely British elements of the UK case have to be identified and allowed for.

The experimental nature of the British model, as signalled in the title of this book, also needs to be remembered, as does the fact that what usually passes for the 'British' model of electricity privatization is that adopted for England and Wales only. Scotland and Northern Ireland had their own separate ESI privatizations and had to make do with monopoly models rather than competition models due to their different system sizes and configurations, histories and political circumstances.

Our objective, therefore, is to explain the structure and regulation of the privatized ESI, and to analyse its performance not just in narrow financial terms but more broadly in terms of economic and distributional effects, and its impacts on the British primary fuel industries.

APPROACH

This book is an independent assessment, not a piece of advocacy either for or against electricity privatization, liberalization and competition. The fact that much of our analysis is sceptical and draws attention to unresolved issues rather than to future opportunities does not exclude the possibility that considerable net benefits may yet arise. We repeatedly identify the importance of the very lenient initial settlement which was determined by the Government (as opposed to the regulator) as the prime cause of the imbalance between the benefits received so far by shareholders on the one hand and consumers on the other. Our sceptical and measured approach is also something of a reaction to the often exaggerated claims made by enthusiasts for privatization, liberalization and competition.

Our approach has also been guided by four basic points of view. Firstly, the privatization of the UK ESI can only be understood by reference to the political and economic circumstances of the time and the place in which it occurred. The context here is everything. Secondly, ESI privatization should emphatically not be treated purely as an act of economic management driven by a particular theory of competition, important though the new theoretical underpinning was to the political ideology. Because of the great importance of this industry to the country, there is bound to be a complex interaction with political factors, and many of the decisions will be taken in the light of pragmatic advice from civil servants and consultants as opposed to advice tendered on theoretical or ideological grounds. Thirdly, changes in the nature of commercial risks within the ESI and in the relationship between the industry and its customers and its fuel and plant suppliers can have profound long-term strategic effects. Fourthly, a process of restructuring, liberalization of the purchasing policies of the electric utilities and the promotion of competition will inevitably be one of extended evolution. The previous public ownership model was still evolving after four decades. There can be no once-and-for-all settlement. There will always be an ongoing agenda of both substantive policy issues and regulatory issues.

In addition, we have strenuously sought to separate the effects of electricity privatization as such from what could have happened had privatization not occurred. For instance, reduced fuel costs of generation would have occurred anyway and therefore cannot be attributed wholly to privatization. Whether they would have translated into the same price reductions is speculative and depends on whether the Government, under the assumption of public ownership being retained, would have kept far more of the economic rent for itself. Also, many of the job losses in the British coal industry would have occurred anyway due to reduced demand levels, de-stocking at power stations, and the great productivity growth in the mines due to rationalization, technical change, and the multiskilling of mineworkers which, among other things, reduced machine downtime and the number of jobs. Activity in the renewable energy sector has clearly been stimulated by the subsidy and protection available since the Government decided to extend similar support to renewables as was available to nuclear power (although the total size of the subsidy going to nuclear power was vastly greater than that for renewables). But who is to know what other schemes might have been adopted to promote renewables had the ESI remained in public ownership and the Government entered into the same agreements at the 1992 Rio UN Conference on Environment and Development?

Many of the changes which have occurred in the period since ESI privatization could equally well have occurred if public ownership had been retained and liberalization measures had been introduced. These possibilities include allowing gas, imported coal, and Orimulsion (a liquid fuel based on Venezuelan shale oil) to be used for power generation; allowing the interconnector with France to be used solely for imports; and lifting the de facto ban on importing heavy electrical plant and equipment. All these were administratively and technically feasible had the ESI remained under public ownership, but the political thrust was for privatization, as opposed to liberalization within the public sector. The reality was that the Government was not interested in just reforming the old regime.

THE NEW INTELLECTUAL FRAMEWORK

We now turn to the new theoretical underpinning of privatization and competition in electricity supply. The starting point is that the ESI consists of four separable activities:

1 generation (or production);
2 high voltage transport (transmission);
3 local and low voltage transport (distribution); plus
4 retailing and marketing (supply).

Various stages in this chain can be integrated (brought under common ownership or control) or deintegrated, and horizontal integration is also possible (the same company concentrating on any one of the four activities). The predominant changes in the postwar ESI were, until the early 1980s and in some cases beyond, extensions in public ownership and increases in the level of vertical and horizontal integration. Electricité de France is still the classic model of the fully integrated (vertically and horizontally) state-owned electricity utility.

The theoretical justification in economic thinking for these trends was that: firstly, electricity supply constituted a natural monopoly, where average costs fall continuously as output levels rise, with the result that cost minimization can only be achieved by having a single firm at each stage of the chain; secondly, there were economies of integration between the stages of the chain so that overall industry costs would be minimized if there were high degrees of vertical integration; and thirdly, there were large and increasing economies of scale, which demanded large production units.[1]

Some electricity industries remain privately owned (eg in the USA and Japan) but the general trend towards large highly integrated companies is also present in these cases. Substantial monopoly powers mean that these privately owned industries are generally heavily regulated, usually by profit controls (or rate-of-return regulation).

From the early 1980s two related trends have been evident. There has been a revival in economic theorizing about the desirability of market solutions to resource-allocation problems, extending into new spheres (eg health) which were previously thought most efficiently dealt with in political or hierarchical allocation systems. There has also been a political climate, especially in the Anglo-Saxon world, in which privatization and extended use of market mechanisms became the principal means of change. A third trend, shortage of capital for infrastructure, especially in newly industrializing countries, also became clear in the 1980s, and this interacted with the first two factors to encourage a growing interest in the privatization of public utility industries. This interest has been especially pronounced in the case of the ESI. For developed countries the interest in introducing competition has tended to be to obtain cost savings in generation (partly by reducing capacity surpluses). For developing countries it has tended to be the need to have a framework of partial or full privatization (and profit repatriation) to secure investment funding from international capital markets to support economic and social development. Notice that the latter interest is not necessarily compatible with competition: international investors are more likely to be interested in monopoly solutions to guarantee acceptably high levels of profits on their investment than in competition as such.

Recent economic thinking as applied to the ESI has concluded that generation and retail supply are not natural monopolies. In other words there are no major cost advantages to ever-increasing scale in these activities. Given the presumption in orthodox micro-economics in favour of competitive markets over monopoly, on the grounds that competitive industries will control costs better and produce more output at lower prices than monopolists, strong arguments have been made for competition at the generation and supply ends of the market. Once these arguments are made for horizontal de-integration, the case for vertical de-integration also becomes strong. This is because vertically integrated companies controlling part or all of the monopoly network are likely to use the market power of that monopoly to favour their own generation or supply businesses. Such cross-subsidies and anti-competitive behaviour will frustrate the development of real competition. In addition, other market economics work on the boundaries of the firm suggests that in most situations total costs will be minimized by firms sticking to core businesses and contracting-in other services on competitive terms from the market.

Together these arguments represent a strong challenge to the old presumption in favour of integrated monopolies. But they do not prove that competitive, de-integrated solutions will always minimize costs. Electricity is an unusual commodity, particularly in the interdependence between stages of production – it is no use building new generation capacity if there is no transmission or distribution capacity to carry it to customers – and the fact that it is effectively non-storable raises particular problems of system coordination. Economies of scale may still have some significance at the generation and supply ends of the market. These arguments suggest that there may also be significant economies from integration – both horizontal and vertical – that need to be set against the probable gains from competitive behaviour. Put another way, the *absence* of integration may impose high transaction costs on the various firms coming to the contractual arrangements necessary in a de-integrated market. Whether these transaction costs outweigh the competitive gains is then an empirical matter to be determined by research.

Efficiency in electricity supply may therefore best be promoted by competition, or by integration, or in practice by some judicious blend of the two. Although competition may be effective as a means to lowest costs in generation, the potentially very high transaction costs (eg in trading and settlement systems, new metering, and arrangements to guarantee security of supply) may outweigh efficiency gains that may be available at the smaller end of the supply market. This book does not make any a priori assumption about the presumed superiority of either of the general approaches to efficiency outlined here, and in so doing, recognizes that the competitive argument is not axiomatically the superior path to higher efficiency.

International Developments

Many countries besides the UK are now moving towards either liberalization or privatization, or both, in economic organization generally and for the electricity supply industry in particular. Examples are the (slow) movement towards a free internal market in electricity in the European Union, the attempt to restructure and introduce competition in generation and supply in parts of the USA, the virtually wholesale privatization going on in Latin American electricity systems, and the market-based experiments in some East European countries; there are also large injections of private capital into new investment in many Asian power systems. The reform process is highly variable across different countries.

The British experiment is of particular interest to the rest of the world for three reasons. Firstly, it occurred relatively early in the international reform process. Britain was not by any means first (Chile for instance provides significantly earlier experience of privatization), but Britain was the first major industrialized country to embark on radical electricity reform. Secondly, the British model has been comprehensively applied in England and Wales. There are several possible avenues for reform: privatization (change in ownership); vertical de-integration; horizontal de-integration and competition in generation; competition in supply; and re-regulation. Many countries have attempted some parts of this menu, but only the system of England and Wales has been exposed to the full package and has introduced all the changes simultaneously. Some countries are privatizing (Chile, Argentina, Brazil); others are introducing competition in generation, either across the whole sector (as in Norway, Argentina and the Ukraine) or in parts of it (as in the USA). Countries such as Spain, Portugal and Thailand have introduced new generating companies to supply incremental demand. But very few countries other than Britain and Norway have contemplated supply competition. Many countries are introducing new methods of regulation (South Africa, the Scandinavian countries, and parts of the USA). But only Britain has tried to combine all these elements.

Thirdly, the World Bank has considerable influence on power-sector policies in the developing world and it appears to be keen to promulgate the British precedent in order to encourage privatization and market disciplines wherever possible. But the fundamental interest of many developing countries in power-sector reforms stems not from any desire to change the ownership of their power systems but from the fact they have no choice but to go to international private investors if their electricity systems are to grow fast enough to keep pace with demand. The World Bank's influence exposes developing countries to the need to consider more far-reaching reforms than are necessary simply to attract foreign private investment on a large scale. To the extent that this would normally require a virtual guarantee of a good return and profit repatriation, it would tend to point to a monopoly structure to avoid the risks of 'stranded' investment under a competitive structure.

For all these reasons numerous countries are interested in the results of the British experiment in electricity reform, and it is therefore vital to try to distinguish those parts of the process which seem at least potentially exportable, and those which must remain strictly for home consumption. One of our main purposes here is to do this.

THE STRUCTURE OF THE BOOK

The book is organized in four parts. Part I, *From Public to Private Ownership* sets the scene by providing the economic and political context for the British ESI privatization. Chapter 2 outlines the structure, operation and regulation of the ESI when it was in the public sector. Chapter 3 does the same for when the industry was transferred to the private sector, and it also sketches the background of the much wider privatization programme, of which the ESI is 'the jewel'.

Part II, *The Record, 1990–95*, attempts to assess the financial and economic performance of the privatized ESI so far, and its wider impacts. Chapters 4 and 5 set out the development of competition and regulation, the achievements and the problems encountered. The next three chapters consider some of the major consequences: the pronounced fall in the demand for coal, especially British coal, the colliery closure crisis and the huge job losses in the mining communities, and the substantial role in that of the 'dash for gas' which was triggered by ESI privatization (Chapter 6); the incompatibility between an electricity market based on free competition and an unsubsidized nuclear power programme (Chapter 7); and the dependence of renewables generation on subsidy and the market protection which was introduced primarily to assist nuclear power (Chapter 8).

Any change as radical as the British model of privatization almost inevitably creates losers as well as winners. Chapter 9 uses a broad canvas to assess who the winners and losers have been, including more sets of interests than just consumers and shareholders, although these are clearly the major stakeholders.

ESI privatization has thrown up some major unresolved issues and these are covered in Part III, *Key Issues*. Chapter 10 looks at the continuing issues of obtaining competition in generation and in supply, particularly those related to the intended introduction of competition in the domestic (residential) electricity market in 1998 and from the rapid further restructuring following the series of takeover bids which began in 1995. Chapter 11 discusses the various reforms of the system of regulation which have been called for, including reform of the regulators themselves, their public accountability, the basic method of regulation which is price regulation, the possible role for regulation in promoting competition in supply when the domestic market is opened up, and changes which may be needed where RECs are taken over by parent companies with overseas subsidiaries or by other public utilities which operate in the UK. Chapter 12 examines the long-term strategic issues of electricity production and supply, which are easily overlooked when discussion concentrates on competition and regulation: it examines how far the Government's ability to influence outcomes in the public interest has been affected by privatization. The main uncertainty is the ability to 'keep the lights on' under a system without territorial franchises and without an effective mechanism to ensure investment in new capacity.

Part IV, *A Model to Follow?* comprises just Chapter 13, which assesses the various advantages and disadvantages of the British electricity privatization, but avoids drawing up a final balance sheet because it is still far too early to know whether the British model will provide significant price and other advantages to the great bulk of consumers (each with only very small market power) without prejudice to long-term security of supply and universal service provision. It also warns against regarding the British model as one which is universally applicable to the wide range of countries with different conditions and objectives which are adopting or considering privatization as a solution for the problems of their own power sectors.

2

UK Electricity Supply under Public Ownership

John Chesshire

This chapter describes the evolution, structure and main functions of the UK electricity supply industry (ESI) primarily under some forty years of public ownership until 1989, and examines ways in which many key policy issues were addressed. Its aim is not to provide a detailed history but rather a context to assist understanding of both the process and the early experience of privatization.

Structure and Duties of the Nationalized Industry

Public electricity supply in the UK dates from 1881 when Siemens began operation of a small hydro-electric generating plant in Godalming, Surrey. This followed a series of small-scale experiments of electric lighting in various towns in the UK and in Paris and New York. The Godalming company lasted but a short while, ceasing operations in 1894. The next UK public supply was established in Brighton on 27 February 1882 and, as a result, Brighton has a very strong claim to have experienced the world's longest continuous public electricity supply.[1]

From its foundation, to nationalization in the 1940s, the UK ESI was based upon small-scale, local, private or municipal companies although there were successive series of mergers and re-alignments. At the time of the Weir Committee Report of 1925 there were some 600 separate electricity supply undertakings based upon over 400 generating stations.[2] Following the Committee's recommendations, the Central Electricity Board (CEB) was established in 1926 with responsibility for constructing a high-voltage national electricity grid network. The CEB operated as a statutory corporation, similar to the BBC, rather than as a nationalized industry. This was the first attempt to create a national executive body capable of integrating disparate local supply networks. The Electricity (Supply) Act of 1926 mandated the CEB to embark upon standardization of the frequency of alternating current across the UK so that interconnections could be made, and also to oversee the planning and construction of new generating capacity. The CEB developed a grid control system and encouraged construction of larger capacity and more thermally efficient generating plant. This was to be a long-term programme: even in 1947, immediately prior to nationalization, only two-fifths of the 569 distribution undertakings were being supplied directly by the grid. Even then power supply remained essentially a local activity, with only limited power transfers between these small utilities.[3] As a result of joint planning and increased integration, the plant reserve margin was reduced from 84 per cent in 1929–30 to 16 per cent in 1939 leading to improved system reliability and higher capital productivity and system thermal efficiency.[4]

The ESI in England and Wales (and Southern Scotland) was nationalized in 1947 as the British Electricity Authority (BEA). The structure then comprised the BEA – responsible for the generation and bulk transmission of electricity – and 14 statutory independent Area Electricity Boards (12 AEBs in England and Wales and two in Southern Scotland – responsible for local distribution, metering, billing, customer advice and ancillary activities such as contracting and appliance sales in their shops). At inception the BEA operated 297 power stations with a total generating capacity of 12,900 MW.

The North of Scotland Hydro-Electric Board (NSHEB) was established by statute in 1943 with responsibilities not only for public electricity supply and exploitation of hydro-electric potential, but also with a central role in the economic development of the remote and sparsely populated areas of Northern Scotland. In 1955 the South of Scotland Electricity Board (SSEB) was created to serve the more industrialized and heavily populated areas of southern Scotland and given additional responsibilities for generation and transmission. As a result, both Scottish Boards were now autonomous and responsible for generation, transmission, distribution and supply in their respective service areas. From 1965 the Scottish Boards operated a joint generating account, the coverage of which was further refined in 1977. This account permitted joint planning and operation of generating plant on the Scottish mainland which was ranked in a merit order, and enabled differences in tariffs between the Boards to be kept within small limits.[5]

The main features of the UK ESI in terms of generating capacity, plant mix and electricity sales are summarized in Table 2.1.

Table 2.1 Main features of UK ESI, 1987

Capacity by type	*Capacity (MW)*	*Electricity generated (GWh)*	*Consumer category*	*Number (thousands)*	*Sales (GWh)*
Fossil fuel	50,263	226,382	Domestic	22,383	93,254
Nuclear	6,519	50,282	Farm	263	4,109
Hydro[a]	4,085	3,312	Industrial/ commercial	2,103	153,689
Other[b]	3,001	512	Other[c]	5	5,137
Total	63,868	280,488	Total	24,754	256,189

Source: Handbook of Electricity Supply Statistics, 1988, Electricity Council, pp 132–133.
Notes: (a) Hydro includes hydro and pumped storage plant.
(b) Other capacity includes single-cycle gas turbines and internal combustion engines.
(c) Other consumers includes public services, street lighting etc.

Like the two Scottish Boards, Northern Ireland Electricity (established by statute in 1972) operated as a vertically integrated utility with responsibility for generation, transmission, distribution and supply. In 1970 a 300 MW interconnector commenced operation with the Electricity Supply Board in the Republic of Ireland; this was destroyed by terrorist action in 1975 and was not re-commissioned until April 1995.[6]

In 1955, with the creation of the SSEB, the BEA became the Central Electricity Authority. It was a short-lived re-organization. Following the Herbert Report,[7] the 1957 Electricity Act further amended the structure in England and Wales by creating the Central Electricity Generating Board (CEGB) responsible for power station operation and construction and for the national grid; and the Electricity Council with broad supervisory responsibilities for the industry in England and Wales. The 12 AEBs in England and Wales were left in place essentially as retail distributors of electricity.

Due to problems which had become evident by the late 1960s – particularly the fact that the CEGB generated the bulk of the industry's costs and passed them on automatically to the Area Boards through the Bulk Supply Tariff (see below) and was thus insulated from many cost disciplines – two unsuccessful efforts were made to restructure the industry in 1969–70 and again in 1976–77. One proposal, under discussion in Whitehall in the late 1960s, was to create perhaps four vertically integrated regional boards for England and Wales with the aim of stimulating competitive performance and of weakening the CEGB's dominance over investment and technology choice.[8] Re-examining the same issues in 1976, the Plowden Report proposed a stronger, unified Electricity Corporation to be charged with determining strategic policy for the ESI as a whole in England and Wales, with subsidiary operating boards similar to those of the CEGB and Area Boards.[9] In effect, therefore, *strategic* decisions were to be taken from the CEGB which would have remained responsible for building and operating power stations and the national grid. These proposals both failed due to a shortage of Parliamentary time and because the proponents of change (the Labour Party) lost the general elections in 1970 and 1979.

The ESI was by far the largest of the UK's nationalized industries in terms of turnover and capital employed. By the late 1980s the combined net assets of the ESI exceeded £42 billion, including the CEGB (£27 bn), the AEBs (£10 bn) and the Scottish boards (£5.6 bn). In 1989, immediately prior to privatization, the structure of the UK ESI comprised 17 statutorily independent boards with varying functions.[10] In Northern Ireland and Scotland, respectively, the NIE, NSHEB and SSEB operated as vertically integrated utilities responsible for generation, transmission, distribution, supply and ancillary functions such as electrical contracting, street lighting and retail outlets for electrical appliances in their service areas.

In England and Wales there was a federal structure in which the CEGB supplied 95 per cent of power requirements in England and Wales, the remainder comprising numerous small, industrial self-generation plants; the nuclear stations owned by the UK Atomic Energy Authority (UKAEA) and British Nuclear Fuels Ltd (BNFL); and imports from Electricité de

France (EDF). The CEGB operated the high-voltage national grid (above 132 kV) including links with France and Scotland; and designed and ordered power stations. It had a statutory obligation to meet all bulk supplies of electricity required by the AEBs in England and Wales. The CEGB supplied only very few customers directly, such as British Rail, the UKAEA and an aluminium smelter on Anglesey in North Wales.

The 12 independent Area Boards were responsible for the distribution and supply of electricity from the national grid offtake points to final customers in their areas via their 132 kV and lower voltage distribution networks. The Boards had a statutory duty to plan and operate the distribution and supply of electricity to all of their customers; they also operated appliance showrooms, undertook electrical contracting and servicing, and other functions such as street lighting. After 1957 the Boards had the power to generate electricity subject to consultation with the CEGB, Electricity Council and Government, although the Midlands and the South Western Electricity Boards were the only ones ever to operate a few, small generating schemes.

The Electricity Council coordinated industry-wide policies and was responsible for central management of finance, taxation, industrial relations, R&D, national advertising and marketing campaigns and for advice to the Government. Crucially the Council had no control over the CEGB or the Area Boards, and the latter had little influence over the CEGB.

Limited transfers of electricity occurred between the CEGB and the SSEB. Both the Scottish and the CEGB generating systems were planned on the basis of zero trade. The original purpose of this interconnection was to provide assistance when supply security was under threat. Capacity was expanded to some 1000 MW in the mid-1960s with this aim in mind. Over the next twenty years, annual power exchanges typically represented 5–9 per cent of SSEB and 1–2 per cent of CEGB energy requirements. Given substantial over-capacity, especially on the SSEB system in the 1970s, pressure mounted for power transfers to be increased on the basis of commercial, not merely security, grounds.[11] These discussions were overtaken by privatization. A sub-sea, surface-laid 160 MW link with France opened in 1961 but suffered damage from trawling. It was taken out of service in 1980 and decommissioned in 1982. A new, trench-based 2000 MW DC link began operation in 1986. Originally conceived to permit balanced two-way flows of power between the CEGB and EDF, this link has been used to import French electricity on an almost continuous basis.

Since 1947 independent, private generation of electricity had declined in importance to about 6 per cent of total electricity production in 1989. Of this, one-third represented production by the UKAEA and BNFL. Little of the privately generated power was 'exported' to the national grid: only about 0.3 per cent of total *public* supplies of electricity.[12] The Government sought to encourage greater competition from independent generators under the terms of the Electricity Act, 1983. This attempt failed given: the use of a much lower discount rate (5 per cent real) by the CEGB than available to private generators; the CEGB's moves to realign fixed and variable costs in the Bulk Supply Tariff; and the lack of independent arbitration of disputes over electricity prices and transmission charges. A prime reason for the displacement of private by public supply electricity over this period was because of higher capital and thermal efficiency in bulk power generation and transmission. Between 1937 and 1987, electricity prices increased by a factor of only 9, whereas fuel costs and the cost of living (Retail Price Index) rose by factors of 55 and 21 respectively.[13]

The main statutory obligation of the Boards in the nationalized ESI was to develop and maintain an efficient, coordinated, economical and reliable system of supply in their respective service areas. Their other principal duties were to: break even taking one year with another (see below); avoid any undue preference against any consumer, or group of consumers, in setting tariffs or terms for supply of electricity; to promote health, safety, welfare and training of employees; and to have regard to environmental concerns. As electricity generators, the CEGB, NIE, NSHEB and SSEB also had a security of supply obligation which dominated their approach to forecasting, system planning and investment, particularly while electricity demand was growing rapidly. In addition, all the Boards had wider – but unwritten – economic and social obligations (eg they were expected to stimulate other sectors of the economy, principally via their purchasing of fuel and equipment – see below).

Strategic Control of the Industry

Political Control

The original advocates of nationalization proposed that the publicly owned industries should operate in the public interest, guided by a broad statutory framework but at arm's length from Government, with substantial managerial autonomy over operational, investment and planning decisions. Given the scale of resources absorbed by the public sector, and its impact on the wider economy and the fuel and plant supply industries, this vision of arm's-length operation was to prove idealistic rather than realistic. The advocates also anticipated the need for an efficiency audit commission to permit independent monitoring and reporting on performance of the trading public sector. Such a structure was never formally defined or adopted. Rather, it evolved in a haphazard way, split between a range of organizations and always subject to the prior determination of major issues by the Government.

Over time, the Government exercised a growing number of controls over the nationalized industry. Amongst the most important of these was the framework of financial control, discussed separately below. The Government's formal powers derived from two principal sources. General powers were set out in legislation such as the Ministry of Fuel and Power Act, 1945. This required the Secretary of State to 'promote economy in the supply, distribution, use and consumption of fuel and power'.[14] This duty required Ministers to keep under review the security of supply, the resources employed, effects on the balance of payments, and all the economic, social and human consequences of energy policies.[15] Other, more specific, Ministerial powers derived from numerous Electricity Acts and statutory instruments.

These powers were exercised usually by Ministers and officials in the sponsoring Department (eg the Department of Energy, or the Scottish Office) together with the Treasury. The relevant Cabinet Minister of the sponsoring Department appointed the chairmen and full- and part-time board members of each constituent Board. The Minister also approved capital investment and R&D programmes, the latter after prior review by the Advisory Council on Research and Development. The Minister also approved the appointment of auditors, planning consents, overhead lines and wayleaves, meter designs and determination of meter accuracy. In a few cases, the CEGB was required to order generating plant ahead of need to assist employment in the heavy electrical industry; some compensation was given to the CEGB by Governments as a result. As with other parts of the public sector, at times the ESI was required to assist in counter-inflation policy by accepting lower price increases than justified by movements in underlying costs. The ESI, as the only civilian market for nuclear power, was also subject to Government decisions on reactor choice – decisions which were strongly influenced by the UK Atomic Energy Authority acting as the Government's adviser on nuclear power issues. Under Conservative Governments from 1979, the rate of debt repayment was accelerated by use of higher financial targets. As privatization approached, this process of 'fattening up' the industry's profits became increasingly controversial (see below and Chapter 3). As with the gas industry somewhat earlier, the ESI provided a large flow of funds to the Treasury in the late 1980s.

Parliamentary control over the electricity supply, and other nationalized industries, was exercised by occasional Ministerial statements and reports to Parliament (such as via the Minister's Annual Report on Electricity); Parliamentary questions; and scrutiny of the Boards' Annual Reports, finances and policies by Select Committees (that on Nationalized Industries until 1979 and on Energy until 1992). Ministerial and Parliamentary scrutiny was assisted to some extent by the publication of an increasing number of performance indicators, either in annual reports and accounts, or in specific documents. Ad hoc efficiency audits were conducted by a range of bodies, such as the National Board for Prices and Incomes in the 1960s and 1970s, followed by the Price Commission and then by the Monopolies and Mergers Commission. However, the relationship between the nationalized industries, Governments and Parliament was always difficult. When electricity prices were falling in real terms as a result of lower fossil-fuel input prices, technological change and scale economies, the ESI was relatively free from intervention. At other times – when electricity prices rose, when plant commissioning delays mounted, when reactor choice was under review or when large-scale nuclear ordering programmes were proposed – the ESI was subject to close political, media and public scrutiny. Both Ministers and industry managers were often held responsible for situations either not of their making or beyond their power to remedy.

The original nationalization statutes provided for consultative (or consumer) councils, their chairmen serving on the 12 Area Boards and the relevant Boards in Scotland and Northern Ireland. In 1977, to improve consumer representation at national level in England and Wales, the Electricity Consumers' Council (ECC) was created. This comprised the 12 chairmen of the area consultative councils, and other members appointed by the Secretary of State. It became a statutory body in 1983. Tentative proposals for the ECC's chairman to serve on the Electricity Council never reached the statute book. The ECC substantially raised the profile of more conventional consumer issues (eg debt and disconnection); it was also active in more politically sensitive debates over pricing, financial controls, security standards, plant choice and new investment. Surprisingly, the ECC advocated that no such body should continue following privatization. Instead the Regulator appoints local consultative committees, with only a modicum of national coordination. The absence of an effective national consumer voice is now all too apparent, particularly compared with that of the national Gas Consumers' Council which remained in place after the gas industry was privatized in 1986.

FINANCIAL CONTROL

The statutes of the newly nationalized ESI in 1947–48 (as with other trading companies in the public sector) required the Boards to balance their books taking one year with another, after paying interest on borrowings and making due allowance for depreciation. Broadly, the approach adopted was that of average cost pricing. However, particularly given the scale of their investment programmes, pressure increased for a more formalized approach. This was because of the fear that state-owned enterprises were insulated to some extent from economic and commercial disciplines, that returns on investment might be inadequate, and that public sector investment might 'crowd out' investment elsewhere in the economy (whether in the non-trading public sector – such as education, health and roads – or in the private sector). The combined investment programmes of all the nationalized industries were very large, at different times, during the period of public ownership. For the ESI, this was whilst demand (and therefore generating capacity) was doubling every ten years or so, while the National Grid and local distribution systems were being built up, and during nuclear plant construction programmes.

In 1961 the Government introduced a range of financial objectives under which each nationalized industry was set a financial target.[16] This was usually expressed as a rate of return on the whole of an industry's assets, valued under the historic (later current) cost accounting convention. In 1967 the regime was strengthened. The principle of financial targets was endorsed and, in addition, a test discount rate was introduced to apply to all *new* project investment by the nationalized industries. This was set at 8 per cent real in 1967 and raised to 10 per cent real in 1969. The Government also stipulated that their pricing policies should be based on long-run marginal costs (LRMC).[17] In 1978 these broad principles were further refined.[18] The test discount rate used for each project was complemented by the need to meet a required rate of return on an industry's investment programme as a whole (set at 5 per cent real in 1978 and raised to 8 per cent in 1989). Commercial discipline was reinforced by the introduction of performance objectives (or targets), and the application of strict cash limits. Such limits, later termed external financing limits (EFL), defined the change in borrowing permitted from one financial year to the next, either negative (requiring repayment of outstanding debt) or positive (permitting increased borrowing).[19] The industries were also exposed to more stringent efficiency audits by the Monopolies and Mergers Commission (MMC) under the terms of the Competition Act 1980.

There was thus a progression to a more comprehensive, though largely non-statutory, framework of controls. These embraced instruments governing the pricing and financial regime of the nationalized industries, and regular annual reviews of operating performance, investment plans and R&D programmes by the relevant sponsoring Government Department and/or the Treasury (the UK Ministry of Finance). From the late 1960s it was intended that the various financial controls should be mutually consistent and that the major determinant of pricing policies should be estimation of LRMCs. The main four elements, briefly, were as follows.

1 Long Run Marginal Cost Pricing (LRMC)

This is an economic concept which argues that prices in monopoly industries should be determined by the operating and capital costs of meeting additional demand on a continuing basis. It was designed to simulate as far as possible competitive market pricing, recognizing that a perfectly competitive market would tend to generate prices related fairly closely to LRMCs. Conservative Governments after 1979 preferred to pursue 'economic pricing', but this was merely shorthand for pricing according to long-run marginal costs. In the ESI, this pricing rule was subject to elastic interpretation and was also operated with some flexibility. At times of surplus generating capacity, and especially when faced with severe competition from natural gas in space-heating applications, electricity prices tended to reflect short-run marginal costs. When counter-inflation imperatives were dominant, the Government overrode pricing principles. Such breaches of

Treasury-endorsed economic pricing principles tended to be followed by high price increases to catch up with both inflation and LRMCs.

LRMCs proved difficult to calculate, especially given that the National Coal Board (NCB, later called the British Coal Corporation, BCC), which determined 40–50 per cent of the ESI's generating costs, found it very difficult to apply LRMC pricing (as was also the case in the gas industry). LRMCs were usually quantified only in rather vague terms such that, for example, the Select Committee on Energy found it almost impossible to assess to what extent prices were based upon them, or to secure cogent quantification from the Treasury. In the 1980s, pricing was increasingly determined by the Government-dictated pace of debt repayment via the external financing limit (see below). The intended symmetry between the various financial controls then ceased to be relevant in the face of the political imperative of reducing the Public Sector Borrowing Requirement to permit election-winning tax cuts.[20]

2 The Financial Target

This was used as the principal instrument of financial control in the medium term (3–5 years). It was determined following discussion between the ESI and Government and, for the ESI, was usually expressed as a rate of return on net assets valued at historic (and from 1980 at current) cost. In England and Wales each Area Board and the CEGB had their own financial target, consistent with that set for the industry as a whole. This target was set at 1.7 per cent per year for the period 1980/81–1982/83 but was raised successively to 2.75 per cent over the three years 1985/86–1987/88 and to 4.75 per cent for the financial year 1989/90.[21] Similar targets were set for the Boards in Scotland and Northern Ireland.

3 The External Financing Limit (EFL)

This was an increasingly important instrument of short-term control, particularly as Conservative Governments sought to reduce the Public Sector Borrowing Requirement (which includes all nationalized industry borrowing). During the 1970s and 1980s, given large plant surpluses and hence low investment, the ESI was increasingly able to self-finance its investment from depreciation provisions and retained profits. The limit was applied for a single financial year at a time and was intended to complement the medium-term financial targets. If all elements of the financial framework were mutually consistent, then the EFL equalled the difference between gross incoming and gross outgoing flows of cash in the industry. According to the Treasury, a negative EFL was the public sector equivalent of a dividend for an industry with no outstanding debt. In England and Wales, the ESI's EFL was not formally allocated to each constituent Board but contributions were agreed internally with the Electricity Council.

With increased pressure on public expenditure, the ESI was required to meet more demanding financial targets, especially EFLs. The position of individual Boards varied, some being able to clear their debts more rapidly than others. For example, the ESI in England and Wales was set an EFL target of minus £1416 m in FY 1986/87 and of minus £1305 m in FY 1987/88; the EFLs for the two Scottish Boards in FY 1985/86 were minus £230 m for the SSEB and minus £30 m for the NSHEB. The Electricity Council's Medium Term Development Plan for 1983–90 anticipated that the combined debts of the ESI in England and Wales of £4.3 bn in 1983/84 would be fully repaid by 1988/89, with an annual flow of 'dividends' to the Treasury thereafter of some £1.6 bn.[22] The financial impact of the large power station oil-burn during the 1984–85 miners' strike delayed this process somewhat but, at the time of privatization, much of the ESI was essentially debt free.[23]

4 The Performance Objective

The first such objective for the England and Wales ESI as a whole was to reduce controllable unit costs per kWh by 4.25 per cent over the two financial years 1983/84–1984/85, compared with 1982/83, and by 6.1 per cent in 1987/88 compared with 1983/84. Similar objectives were set for the NSHEB, SSEB and NIE. As with other controls, they were gradually tightened in the period immediately prior to privatization. These performance objectives were in most cases more demanding than the roughly analogous 'X' factors in subsequent RPI-X price controls set for the newly privatized industry (see Chapter 5).

Costs and Prices

From the 1970s the broad cost structure of the UK ESI comprised generation with two-thirds of total costs (of which fuel purchases represented some 75 per cent), transmission 10 per cent, distribution 20 per cent and supply 5 per cent. The costs incurred by the CEGB in generating and transmitting bulk supplies of electricity were reflected in the bulk supply tariff (BST), representing 75 per cent of the costs of the 12 Area Boards which fixed retail tariffs for domestic, commercial and industrial customers. Over time, the BST became an increasingly sophisticated pricing mechanism to recover both running charges (or short-run marginal costs) and capacity charges. Running periods embraced peak, standard daytime and off-peak night rates, further refined over time to reflect different periods of each day, as well as working days, weekends and holidays.

The CEGB's capacity charges in the BST did not fully reflect its fixed costs. Instead they served to recover the balance between its total revenue requirement (as determined by the financial target and EFL) and its income from running charges. The CEGB's capacity charges to the AEBs distinguished between peak and basic capacity and were expressed as lump sums. The BST effectively guaranteed the CEGB full recovery of all costs, especially when the fuel cost adjustment clause removed uncertainty over recovery of fuel costs.[24] The BST was thus the mechanism for passing all of the CEGB's costs to the Area Boards and then to final electricity consumers. Similar arrangements for recovering unit- and capacity-related generating and transmission costs were used by the three vertically integrated Boards – the NIE, NSHEB and SSEB.[25]

In England and Wales, market risks were borne mainly by the Area Electricity Boards. Variations in sales forecast by AEBs had little effect on the CEGB's financial position since fuel costs were fully recoverable. Capacity charges also shifted uncertainty on to the AEBs. An Area Board could not be sure until the end of the financial year what capacity charges it had incurred. These depended on the time profile of consumer demand by other AEBs as well as that of its own customers. Also, if Area Board sales were lower than those forecast, average cost per unit would rise in relation to revenue per unit, thus reducing profitability given that tariffs could not be adjusted retrospectively. As with the three vertically integrated electricity Boards operating in Scotland and Northern Ireland, the AEBs developed a range of retail tariffs reflecting customer category, time of day, season, voltage and maximum demand. Within an Area Board, tariffs for a customer group were 'postalized' (eg no distinction was made between rural and urban consumers, even though these incurred different costs). The ESI also failed to reflect fully the cost messages of the BST in AEB tariffs. Field trials with more advanced meters and communication systems were in progress in the 1980s but were overtaken (and perhaps postponed) by the imperatives of privatization.

Planning and Forecasting

Between 1955 and 1970, UK electricity demand grew at an average rate of 7 per cent a year, doubling every ten years. The ESI expected high growth to persist given increased electricity penetration in final demand markets. This process of intensified electrification was, in the UK as in France, expected to permit rapid growth in nuclear generating capacity as a means of reducing heavy dependence on fossil fuels. During the 1970s, the debate over 'energy futures' (embracing forecasts but also plant choice) became increasingly vexed as a range of independent analysts engaged the ESI at conferences and protracted public inquiries.

For much of the postwar period until 1966, the ESI under-estimated electricity demand growth using rather simple extrapolation techniques. Plant shortages were exacerbated by rapid load growth in homes and industry. The UK's flirtation with national economic planning in the 1960s led to high forecasts of GDP growth, in turn used by the ESI to justify high forecasts of electricity demand. In the event neither the economic nor the electricity forecasts proved accurate. From 1970 electricity demand growth fell sharply given lower economic growth, substantial structural change in the UK economy away from heavy industry, and rapid market penetration of natural gas. Significant efforts were made by the ESI to improve medium- and long-term forecasting methodologies, deploying both 'top down' estimates based on overall GDP growth and 'bottom up' forecasts based upon detailed sectoral and final end-use analysis. However, the ESI (especially the CEGB) continued to produce forecasts for electricity demand higher than those of independent analysts; but, by the mid-1980s, the ESI's much-reduced forecasts mirrored those of virtually all other analysts. After very considerable debate the new conventional wisdom had, by then, become one of low electricity demand growth for the UK (1–2 per cent a year).

High forecasts of electricity demand led to substantial surplus capacity, despite delays in plant commissioning. The CEGB's plant margin (an ex-post measure of capacity in excess of maximum annual demand) rose from 21 per cent in 1970–71 to 42 per cent in the period 1973–76. Had the advanced gas-cooled reactors (AGRs) been commissioned on time, the CEGB's plant surplus would have matched that of the SSEB and NIE (well over 70 per cent). From the mid-1970s the ESI engaged in an accelerated plant closure programme as the availability of large fossil stations improved and the AGRs came on stream.

The role of forecasts changed, too, during the 1970s. Until then, forecasts were used to justify investment programmes to satisfy projected peak demand six (later seven) years ahead. To this projected customer demand was added a planning margin to cover for uncertainties stemming from lower availability of plant at winter peak, weather variability and forecasting errors. The CEGB's planning margin rose from 14 to 17 per cent in 1964; to 20 per cent in 1970; and to 28 per cent in 1976–77. There was considerable anxiety about the resource costs of such a large planning margin on the part of Government, consumer bodies and some independent analysts. The planning margin and the generation security standard (the probability of voltage or frequency variations and of load loss) were subject to frequent review.[26] However, at the time the 28 per cent planning margin was adopted by the CEGB, its role in influencing investment decisions had become almost wholly unimportant.

This was because the CEGB began to use a very different approach to investment appraisal. This approach simulated lifetime operation of potential new plant on its system against a given 'planning background', built up of load forecasts, relative fuel price assumptions, capital costs and the progressive retirement of existing plant over time. Alternative new plant options were then assessed against this background to calculate the annuitized net present costs of each option on the system as a whole. These costs were expressed in £/kW per year, and called the net effective cost (NEC); a similar approach was used to estimate the net avoidable cost of retaining old plants on the system. If a project assessment produced a negative NEC it was judged worth undertaking, as the lifetime annuitized savings in fuel and operating costs were calculated to outweigh all construction and financing costs.

In a generating system with a plant mix far removed from the optimum, given the historically accumulated capital stock which reflected dramatic changes in relative fuel prices, such an approach to new investment had numerous advantages of transparency and rigour. It did not obviate the need for difficult judgments about key economic and technical assumptions (eg absolute and relative fuel prices, construction lead times, lifetime operating performance). The CEGB used this methodology to support its long-term development strategy, set out in the Board's 1979–80

Development Review at the time of the second oil shock. The key elements of the strategy were: '(i) developing nuclear power; (ii) containing costs and (iii) diversifying fuel sources to avoid excessive dependence on one single source'.[27] The Development Review concluded that:

> nuclear plant retains its clear advantage over all other competing types of plant. This is expected to persist until after total nuclear capacity on the system exceeds the system base load requirement, possibly beyond the year 2000.[27]

The timing of this strategy coincided with the newly-elected Conservative Government's determination in the aftermath of the 1978–79 oil shock to launch a large pressurized water reactor (PWR) nuclear power programme, partly to diversify fuel supplies, partly as a check upon the industrial and political power of the mining unions. The CEGB's programme, endorsed in principle by the Government, envisaged construction of one new nuclear station a year in the decade from 1982, or some 15 GW over ten years.[28]

The NEC approach meant that new nuclear plants could be ordered, despite a large plant surplus, if it could be shown that the NECs of a tranche of nuclear plants were negative. The use of a 5 per cent (real) discount rate crucially assisted a capital-intensive, long lead-time technology such as nuclear power. The increasingly pro-nuclear stance of the CEGB, and the assumptions it deployed in its forecasts and economic analyses, led to intense debate amongst specialists and 'anti-nuclear' environmental groups. Events such as the protracted Sizewell B and Hinkley Point C Public Inquiries were major occasions. That for Sizewell B was the longest planning inquiry in the world, commencing in January 1983, concluding in March 1985, with the Inspector's report presented to the Government in December 1986. Whilst the report favoured construction of the Sizewell B PWR plant, during this inquiry fortune turned sharply against nuclear power given the Chernobyl nuclear accident (1986), low international oil and coal prices from early 1986, and defeat of the industrial and political power of the coal-mining unions after the bitter dispute of 1984–85.

The combined nuclear enthusiasm of the Government and the CEGB were also steadily eroded by a series of other major investigations into nuclear power in the 1980s. Perhaps the most telling investigations were those by the MMC, the National Audit Office and the Select Committee on Energy. By the time of privatization, public knowledge about nuclear costs, decommissioning liabilities, dubious forecasting and capital investment appraisal methodologies was substantial; even once-staunch supporters of nuclear power were now very sceptical.[29] Two events reveal the reversal in nuclear power's fortunes over the decade following 1979: the ordering of but one of the intended large programme of PWR nuclear plants (Sizewell B, commissioned in 1995); and the failure to convince the City and public of the attractiveness of privatized nuclear plants. In July 1989, the Government withdrew Magnox plants from the privatization, followed by all other nuclear plants in November 1989 (see Chapter 7).

PURCHASING AND R&D

FUEL PURCHASING

Until the early 1960s, the ESI's fuel requirements were met largely by coal, together with small contributions from oil and hydro-electricity. Given forecast growth in electricity demand and fears of a shortfall in indigenous coal production, the CEGB began converting some existing stations to oil and ordering new oil-fired plant. In addition, in 1955 the ESI announced the world's first commercial nuclear power programme based on Magnox plant. This was based on eight stations comprising 3427 MW. A second programme of 6480 MW based on the advanced gas-cooled reactor (AGR) began in 1964/65. Two further AGR stations were ordered in the late 1970s, pending design work on the PWR. One 1200 MW PWR station (Sizewell B) was approved by the Government in 1987, shortly after the planning inquiry inspector presented his report to the Minister, and this plant was commissioned in 1995.

UK hydro-electric capacity was limited by topography, but significant pumped storage capacity was developed at Ffestiniog and Dinorwig in Wales. A small amount of natural gas was consumed in the early 1970s, largely to assist the gas industry until load developed in final markets. Until privatization, the use of gas in power generation was never very large given that UK Governments considered natural gas as a 'premium' fuel best consumed directly by final consumers. All gas was purchased from the British Gas Corporation, which took a similar view about such gas use. Gas used in power generation peaked at some 4 mtce in 1975, but subsequently fell to below 0.5 mtce annually. The CEGB ordered some single-cycle gas turbines, based on light oil, in the 1960s and 1970s. They were used largely to provide local security on remoter parts of the grid, and to permit emergency start-up of main generating plant in the event of grid system failure.

In the 1980s, the CEGB took tentative steps to invest in small wind power projects, essentially to become familiar with the technology. It was criticized for failing to support renewable sources and combined heat and power (CHP), and for its predilection for large plants in general and for nuclear power plants in particular. At the time of privatization, the industry had no firm plans to order gas-fired combined cycle gas turbines (CCGTs), subsequently to become the preferred generating technology for the privatized utilities. Rather the CEGB was planning to order a new vintage of coal-fired stations with twin 900 MW units, located on the coast to permit use of imported coal, and four more PWRs. The change in attitude of the UK Government, and the European Commission, to gas use in power generation – combined with the considerably increased production on the UK Continental Shelf, which led to much lower beach (or wholesale) prices of gas – was essentially contemporaneous with the privatization of much of the ESI and a very important factor in its early 'dash for gas' (see Chapter 6).

Despite this diversification of primary fuel sources, the UK electricity industry remained heavily coal-dependent. By the late 1980s, the CEGB met some 75 per cent of its fuel requirements from coal, and 20 per cent from nuclear power. Coal purchases represented some 50 per cent of the CEGB's total costs, and some 40 per cent of the final electricity price. All other fuels and technologies played a minor role. As the largest purchaser of British coal, the CEGB was in a position of bilateral monopoly: the NCB/BCC supplied virtually all the coal required by power stations and was easily the largest fuel supplier. Coal prices and volumes were therefore largely determined by periodic hard-fought bargaining. At no time was there a long-term coal contract because the NCB/BCC was unable to guarantee prices and the CEGB did not want to guarantee long-term quantities.

Given the strategic importance of both industries, it was inevitable that UK Governments considered it essential to oversee the bargaining process. Other than at times of acute supply shortfalls (eg the mid-1950s), the UK ESI purchased virtually all of its coal from indigenous sources. This led to its strategic vulnerability when the National Union of Mineworkers became increasingly active from the 1970s, and took strike action – especially the bitter and prolonged dispute of 1984–85.

The CEGB's first purchase of foreign coal was authorized by the Government in 1970. Occasionally UK Governments sought to exercise explicit control over coal imports, as in 1976 (when sterling came under intense pressure). At other times this option was implicitly or explicitly rejected by the CEGB, as the Board's judgement was that it was unlikely to be permitted to purchase significant volumes of imported coal. From 1979 the CEGB no longer required Government approval to import coal. With annual sea-borne steam coal volumes in the range 100–140 m tonnes in the 1970s and 1980s, the CEGB judged that any significant switch to imported coal (say above 10 m tonnes a year) would have led to sharp upward movement in international coal prices. In addition, internationally traded coal was priced in US dollars and, for much of the period before the growth in UK offshore oil production from 1975, British economic policy aimed to prevent further devaluation of sterling against the dollar.

The CEGB thus sought to apply greater discipline on the NCB/BCC from 1979, via a series of medium-term joint Understandings. Though not legally enforceable, these committed both parties on a 'best endeavours' basis. The Understanding of 1979 committed the NCB to supply 75 m tonnes of coal annually to the CEGB; in turn the latter agreed to purchase this volume subject to the average coal price not rising in real terms. Later periodic renegotiations of the Understanding in 1983, and subsequently, disaggregated volumes into tranches, increasingly related to movements in international coal prices.[30] Transitional arrangements for coal purchase at the time of privatization continued this process, but with even stronger arm-twisting by the Government to secure agreement between BCC and the generators as a precondition of a successful flotation of the electricity industry (see Chapter 3).

The UK ESI sited the bulk of its oil-fired capacity on the coast, usually adjacent to major refineries, to minimize transport costs and to secure keen prices. With the exception of Fawley, the CEGB always ensured more than one fuel supplier at oil-fired stations. It also staggered contract dates, stipulated maximum and minimum quantities for oil supply and supplemented contract volumes with spot market purchases to minimize exposure in the volatile international oil market.

As regards nuclear fuels, the CEGB and SSEB (the UK's two nuclear generators) were dependent upon the monopoly services provided by BNFL for fuel fabrication, enrichment and reprocessing. Uranium ore concentrate for the UK was purchased by the UKAEA until 1974, from BNFL until 1979 and thereafter from the British Civil Uranium

Procurement Directorate. Civil uranium supplies were obtained from Canada and Namibia. Until 1978, the CEGB obtained enriched uranium from the UKAEA's plant at Capenhurst. In 1971, the UK, Netherlands and West Germany agreed development via URENCO of centrifuge enrichment based at Capenhurst and Almelo. The CEGB and SSEB agreed to purchase virtually all enrichment services from URENCO, BNFL being the dominant UK partner. Much secrecy surrounded the financial terms of the contracts between BNFL, the CEGB and the SSEB but the services provided by BNFL included enrichment, fuel fabrication, irradiated fuel reprocessing, plutonium separation and nuclear waste management.

As with coal, so too with nuclear power. The CEGB was initially hostile to a nuclear power programme in the 1950s. It became inevitably dependent on monopoly supplies of nuclear fuel and fuel services. For many years until the late 1970s, UK Governments remained firmly wedded to British nuclear technology which also made the CEGB and SSEB dependent upon the UKAEA for nuclear R&D, reactor design and safety analysis. The CEGB and SSEB were also dependent upon the British nuclear plant supply industry which remained fragmented, under-resourced and technologically weak for much of the period from the mid-1950s until privatization. Correctly criticized for numerous failings, it must nevertheless be remembered that the CEGB's costs and performance were heavily influenced by the 'buy British' regime for fuels and equipment.

Equipment Purchasing, Commissioning and Performance

During the Second World War, little generating plant was ordered in the UK; much was old and inefficient or had been damaged by bombing. Serious power shortages threatened postwar economic recovery. A statutory order in 1947 limited generators to standardized 30 and 60 MW units to assist much-needed capacity expansion. Rapid scaling-up of set sizes was to occur a decade later: 100 MW sets were commissioned in 1955–56; 120 MW and 200 MW sets in 1959–63; followed by a few units of 275–375 MW. No less than 51 500 MW units entered service between 1966 and 1982; and the first of a series of 660 MW units in 1976. In the period 1960–75 46 GW of new generating plant was commissioned in the UK, 32 GW of which was coal-fired. Facing problems with AGR plants and industrial relations problems in the coal industry (in 1971–72 and 1973–74), the ESI ordered 9460 MW of oil-fired plant in the early 1970s. These plants included Grain, Ince B and Littlebrook D (CEGB), Inverkip (SSEB) and Kilroot (NIE). In hindsight it was a misfortune that large-scale ordering of such plant coincided with the first oil shock in 1973–74. Despite various calls for these plants to be converted to coal, they were completed as a means of diversifying the fuel mix. Most of the CEGB's oil-fired plant subsequently operated at low load factors except during the 1984–85 miners' strike. It seems inconceivable that the decisions by the CEGB, SSEB and NIE to complete the large oil-fired station programme did not receive Government approval.

The national grid, originally rated at 132 kV, was reinforced by a 275 kV supergrid from 1953 and a 400 kV supergrid from 1966. The main strategic function of the supergrid was to permit bulk transfers of power from the coal-fired stations built in the coal-mining areas of Yorkshire and the North Midlands to the major population and load centres in the South East, West Midlands and North West. Investment in distribution by the Area Boards, NIE, NSHEB and SSEB also rose steeply in the 1950s and 1960s to keep pace with demand growth. Virtually all generating plant and transmission and distribution equipment was ordered from UK manufacturers.

Given its comparative size, the CEGB dictated the pace of change in generating technology. This technological dominance of the CEGB was reinforced and partly caused by the absence in the UK of firms able to construct plant on a turn-key basis, as in the USA; and by the CEGB's decision in 1958 to internalize most of the mechanical and electrical engineering required for plant design and construction in its own project groups; and to rely little on external engineering consultants save for civil engineering.

Failure to rationalize plant suppliers, given belief in the virtues of competitive tendering, led to a bewildering array of equipment supplied by eight major boilermaking, and five turbine generator, firms in the 1950s: virtually no two stations had the same combination of boiler and turbine maker.[31] The large increase in plant ordering, combined with rapid technological change, overwhelmed the technical resources of the plant suppliers. Step-jump increases in set sizes, with little time to incorporate learning from early operation, led to reduced availability and delays in plant commissioning. These delays became acute in the early–mid-1960s, leading to severe plant shortages, and a major inquiry.[32] Similar delays with the AGR nuclear programme were experienced in the 1970s. The reasons were also similar: rapid scaling-up of plant (from a 30 MW AGR prototype to

660 MW units); and a preference for competitive equipment supply from a range of small, under-resourced nuclear plant manufacturers.

There were probably several reasons for the rapid scaling-up of fossil and nuclear plant size. These included fears about increasing difficulty in obtaining planning consents for new sites and thus choice of larger units to conserve available sites; and general confidence among power plant engineers in various countries that generation would be subject to increasing scale economies in set sizes up to 1000 MW or even 1500 MW.

Rationalization of plant suppliers, much-reduced ordering, better on-site industrial relations and tighter management disciplines resolved many of these problems with the much-diminished number of conventional fossil-fuel plants ordered in the 1970s. Perhaps for too long, and with Government support, in the 1970s the CEGB favoured the notionally competitive approach of supporting two major British suppliers of key items of conventional plant (which appeared to reconcile the policy of national procurement with that of competition). To maintain this duopoly structure, the CEGB awarded orders alternatively to the two manufacturers (on the basis of 'Buggin's turn'). Substantial cross-border rationalization in the world heavy electrical industry, combined with the European Union's procurement policies, would have forced a change in the nature of customer/supplier relationships upon the CEGB had the ESI remained in public ownership. Poor AGR plant performance continued until privatization. Improved AGR performance was only to be achieved in the 1990s – a long time after the AGRs ordered from 1966 to 1973 came into operation. (See Chapter 7).

Research and Development

For the first decade of public ownership, with unit sizes stabilized at 30 and 60 MW, research undertaken by the ESI was largely confined to work on high voltage distribution and transmission. In 1956 the Herbert Report recommended higher R&D expenditure. Over time, the industry's R&D programme grew substantially, covering generation, utilization, the environment, and fundamental science (such as cryogenics, magneto-hydrodynamics and various combustion and emission abatement technologies). With the exception of EDF in France and ENEL in Italy, this large-scale R&D commitment was rare amongst the world's electric utilities and, for a long time, the CEGB was seen as a world leader. In other major OECD countries, with a more fragmented utility structure, most R&D was left to the equipment suppliers or more recently to collaborative ventures such as the Electric Power Research Institute (EPRI) in the USA.

As the scale of the industry's R&D expenditure rose, criticisms were voiced. Some of the criticism centred on the CEGB's large expenditures on 'exotic' technologies (eg magneto-hydrodynamics). Other critics argued that the CEGB's heavy involvement in generation research weakened the plant suppliers' own programmes. Focusing on larger set sizes was thought likely to weaken manufacturers' capability in export markets. Conversely, given the fragmentation of the UK heavy electrical industry, the scale of commissioning delays and falling plant availability, and its security of supply obligations, the CEGB had little choice but to seek strategic oversight of the plant programme. It was thus forced much beyond the point of being merely an 'informed buyer'. When serious design faults became apparent with big boilers and generating sets, the CEGB devoted a large proportion of its technical resources to their resolution. Once these problems were overcome, from the late 1970s, the CEGB made strenuous efforts to shift much of the R&D burden on conventional plant back to the heavy electrical industry.[33] By then, faced with a dearth of orders for generating plant, the equipment suppliers were reluctant to increase their own R&D expenditures.

As the CEGB's commitment to nuclear power grew, so did its expenditure on nuclear R&D. The largest component of nuclear R&D, on Magnox, AGR, SGHWR, HTR and the fast reactor, was undertaken by the UKAEA. When the Government and CEGB adopted the Pressurized Water Reactor (PWR) in 1979, the CEGB spent large sums to become familiar with this technology in all its aspects (especially the Westinghouse design codes and safety tolerances) and to adapt it to UK safety requirements.

Latterly a hesitant CEGB was forced to expand R&D on the environment, particularly to address issues such as acid rain. It was also frequently criticized for failing to commit sufficient resources to renewable energy sources. Mainly via the Electricity Council, the ESI also engaged in utilization research aimed primarily at enhancing electricity's market share in competitive energy markets (eg electro-heating technologies for industrial process applications) or in flattening the load curve (eg off-peak storage heating). The industry frequently complained of the difficulties it faced in securing interest from UK companies in developing technologies demonstrated in the ESI's laboratories. As widely expected, privatization led to very substantial reductions in R&D funded by the UK ESI, including the closure of most of the industry's large and highly regarded laboratories.

Some Conclusions

It is not easy to assess the performance of the nationalized ESI in the knowledge of the changes that have occurred since privatization. Assessment of some strategic decisions ascribed to the industry – such as nuclear reactor choice, heavy dependence upon British deep-mined coal and reliance upon British equipment suppliers – is difficult because successive Governments must share equal if not greater responsibility for these. Similarly, the tide of events (and particularly dramatic changes in perceptions about fuel availabilities and relative fuel prices) places those benefiting from hindsight in a hugely advantageous position. For example, had world crude-oil prices remained high following the two oil shocks in the 1970s, the economic logic for use of indigenous coal or for at least some expansion in nuclear capacity would have appeared very much stronger.

The nationalization of the industry following the Second World War was widely seen as essential given the need to restore and expand electricity supply capacity as a precondition for economic recovery. Whilst successive efforts were made to reform the structure of the nationalized industry and to specify more fully its economic and financial objectives, for nearly forty years there was little debate between the major political parties about the basic form of ownership. On this there was broad consensus until successive Conservative Governments launched their privatization programme in the 1980s. Even then, the privatization of the electricity industry was seen as likely to prove perhaps the most politically contentious and technically challenging of them all.

Without appearing too uncritical, the nationalized industry succeeded in expanding capacity and in providing a high quality standard for electricity supply. In international terms the ESI achieved a deservedly high reputation for managerial and technical competence. Given the UK's indigenous energy resource base, and the vicissitudes in international energy markets, the electricity industry played a major role in supporting British coal production. It also provided major market opportunities for the British heavy electrical industry. These were viewed as crucial when the balance of payments was seen as a dominant policy imperative. Its productivity performance, whether measured in terms of fuel efficiency or use of labour, was creditable if not outstanding in world terms. Latterly it attempted to postpone the impact of tighter environmental regulations, particularly to combat acid rain, but this was true of most if not all other utilities around the world.

Consumer complaints about service standards were addressed through a well-established regional, and latterly national, consumer council structure. The improving performance of Area Boards was monitored and compared very carefully in this field; and worst practice was readily identified for corrective action. However, consumers could exert little influence on prices. These were determined by the industry's underlying costs, particularly fossil fuel prices, and increasingly by the financial regime imposed by Governments. Although hesitant in some areas, especially those relating to nuclear costs and demand-forecasting techniques, the industry was surprisingly transparent and a very large amount of information was published. Even the more highly-protected information was capable of reaching the public domain via Select Committee hearings. Outside the USA, few electricity industries were better documented on the public record. Privatization of the ESI has much reduced its transparency, given the all-too-frequent defence of commercial confidentiality.

Structurally, although the UK ESI was federal in character, a principal weakness was the overbearing power of the CEGB in England and Wales in the strategic decisions of the industry. Divorced from final customers, it developed a tendency for 'top down' or 'command and control' decision making. In this context, the countervailing power of the Area Electricity Boards and of the Electricity Council was negligible. Another weakness was the way in which the structure and the tariff-setting hierarchy minimised risk to the industry – especially the CEGB and its power station investment programmes. Consumers bore all the risks of the industry. As the ultimate shareholder, successive Governments were most reluctant to share this burden. The new, privatized regime is thus fundamentally different in character from that which preceded it: shareholders now bear very considerable financial risks.

The nationalized system placed very great emphasis on security of supply, especially the ability to satisfy winter peak demand requirements for electricity. The managerial and political risks of failure to supply were seen to be immense. Together with the minimal risks faced by the industry in its investment decisions, this underpinned the priority given by the nationalized industry to supply security. This crucial – perhaps for an electricity system, the ultimate – test of performance has yet to be faced by the privatized regime now in place. To conclude neutrally and perhaps as yet still sceptically, there is merit in quoting the judgement of the House of Commons Select Committee on Energy following its detailed evaluation of the Government's privatization proposals:

> Reviews of international experience, particularly of the USA and other European countries, do not reveal any strong, or indeed positive, correlation between, on the one hand, utility structure, form of ownership, and the degree of competition, and the level of electricity prices and overall utility performance on the other.[34]

3

The Privatization of the Electricity Supply Industry

Steve Thomas

The privatization of the electricity supply industry should not be seen as an isolated act of government policy. It was the culmination of a concerted policy by the Thatcher Governments over a period of about a decade to reduce the overall level of public ownership and to move decision-making for the productive sector of the economy from public to private hands.[1] This chapter traces the development of the privatization programme and examines the process of privatization of the electricity supply industry, including the reasons for the structure and the mechanisms chosen.

The Rationales for Privatization

Privatization increasingly became the centrepiece of the policy programmes of the Thatcher Governments of the 1980s, and its importance continued with the subsequent Major Governments. Its appeal was based on a fundamental belief amongst the strongest of the Government's supporters that public ownership inevitably led to inefficiencies. There have been disagreements within the Party about how far privatization should go and, for example, at one point a former Conservative Prime Minister, Harold MacMillan, likened further privatization to 'selling the family silver'. Privatization has had various objectives other than transferring assets from the public to the private sector, some stated, some implicit. The importance of these has fluctuated through time and they include the creation of a share-owning democracy, generating revenue for the Treasury and breaking the power of large trades unions. However, it is political antipathy towards public enterprise that sustains the continuing privatization programme.

This antipathy towards public ownership also manifested itself in the sale of much of the stock of housing owned by local authorities (so-called council housing) to the occupiers during the early 1980s. The revenues from this programme did not go directly to central government but their existence did allow central government support to local authorities to be lower than it would otherwise have been and can also be seen as a net benefit to the Treasury.

At the outset, one of the key stated objectives of privatization was the creation of a 'share-owning democracy' and a priority throughout the privatization programme has been that, wherever possible, share sales should be aimed at the general public. This has meant that public share flotations could be contemplated only for very low-risk companies. In most cases, the price at which the shares were sold has proved to be well below the level the shares were subsequently traded at on the Stock Exchange.[2] Privatization has resulted in a large increase in the number of individual shareholders, from about 3 million in 1979 to about 11 million in 1991. However, this has not been sufficient to reverse the long-term downward trend in the proportion of shares held by individuals which has declined from nearly 70 per cent in 1957 to about 20 per cent in 1989.[3] While there have been suggestions that the shares were deliberately under-priced and that the gains in share prices were electoral bribes, the easy money made by the general public increased the popularity of share flotations.

One of the central motives imputed to the privatization programme has been the generation of revenue for the Treasury. In the main revenue-earning phase of privatization from 1984 to the end of the 1980s, the UK government raised a total of £37 bn.[4] The revenue to the Treasury from the privatization of Nuclear Electric and Scottish Nuclear is projected by government to be £3.5 bn. In terms of current overall annual government revenue of about £170 bn, privatization revenues are relatively small sums, but the impact can still be significant. In times of recession, it has allowed public services to be maintained at a level that would have been impossible without politically damaging tax increases. Near elections, it has allowed taxes to be reduced, helping to sustain the Conservative Party's claim to be a party of low taxation. Electricity privatization and other privatizations from the late 1980s, such as that of the water industry, also compensated somewhat for reduced Treasury revenues from North Sea production of oil and gas that followed the world slump in oil prices in 1986. This incentive to maximize Treasury revenues had to be balanced with any objective to give the new shareholders some of the 'rent' from the process. Although it is impossible from outside government to estimate how far generating revenue for the Treasury influenced policy in favour of privatization, there can be no doubting the political value these revenues have had.

Another central element of Conservative Party policy, particularly in the early 1980s was the reduction of the power of large unions. One of the factors behind the defeat of the Labour Party in the 1979 general election was public distaste following the so-called 'Winter of Discontent' when strikes by public-sector unions seriously disrupted Britain. The resulting unpopularity of trades unions gave the Conservative Party scope to introduce reforms to trade union procedures which previously would not have been politically acceptable. In addition to these explicit reforms, privatization also gave the opportunity to divide the power of unions. The existence of large, often monopoly, state-owned companies with strong centralized union representation meant that unions could much more easily disrupt an entire sector of the economy than if the sector was split over a large number of private-sector companies. Breaking up the state-owned organizations into competitive private-sector companies meant that coordination was more difficult. Placing the companies in a competitive market increased the extent to which strike action in one company could damage the prospects of its workforce.

It is difficult to estimate the precise extent to which privatization was adopted to serve interests other than the explicit one of reducing public ownership, but a number of other policy objectives were clearly well served by privatization.

1979–84

In the first five years of the Thatcher Government, from 1979/80 to 1983/84, £2.2 bn was raised by sales of shares in state-owned assets, but all involved companies were already operating in competitive markets. Of this amount, 37 per cent was accounted for by sales of government-owned shares in British Petroleum, 28 per cent by the privatization of Britoil, an oil exploration and production company, and 20 per cent by the privatization of Cable and Wireless, a telecommunications services company.

In some respects, sale of these companies seemed no more than a more concerted continuation of the political 'tennis' that has typified the adversarial two-party British political scene. Postwar Labour Party policy had been that for the basic industries of the country, the 'commanding heights', the state should have some presence so that it could, for example, influence the behaviour of the market, retain and develop key skills and control the use of national assets. This policy was founded on the experience of the inter-war years which saw repeated periods of mass unemployment and low investment. National public ownership of major industries, particularly those based on inherently monopolistic sectors was seen as a way of avoiding this destructive cycle. The Conservative Party was much less sceptical about markets and saw state-ownership as stifling innovation and harbouring out-dated practices in industries which had to compete on international markets. Proposals to nationalize or de-nationalize companies in this sector were therefore common components of opposition party manifestos at general elections since the second World War. Because these companies operated in competitive international markets, issues of restructuring or new regulatory mechanisms were not important.

The cash derived from the sales of state assets was of considerable importance to the Treasury. In the early 1980s, the UK economy was in the deepest recession since the Second World War and the proceeds of these sales were able to mitigate its effects on public sector finances. The profits the public garnered from trading in these shares contributed to feeding the public appetite for regular share flotations.

1984–86

While the first phase might have appeared no more than a continuation of past patterns, this phase marked a sharp break. In the early 1980s, the depth of the recession being experienced in the UK had made the Thatcher Government deeply unpopular. However, victory in the Falklands Islands War led to a wave of national self-confidence that reduced this unpopularity, and the Thatcher Government was re-elected in 1983. This was interpreted by the Conservatives as a vindication of the more radical policies of the Government and, with most of the traditional targets for privatization already picked off, the signal to embark on a new phase, this time concentrating on the utilities.

Despite the Conservative Party's traditional antipathy towards public ownership, the state ownership of the utilities had not been seriously challenged since their nationalization. The first target, British Telecom, was well chosen. While telecommunications was clearly a vital area of the economy, its provision to private citizens did not raise the same social issues and political risks as did other utilities. Possession of a telephone was not seen as a basic need and right in the way the provision of water, fuel and power was. The recently completed break-up of the monopoly of the US telecoms giant, AT&T gave a precedent that the break-up of national monopoly in telecoms was not only feasible, but could be a positive measure to encourage innovation and entrepreneurship. Telecoms was also seen as a fast-growing, technologically dynamic sector in which major new products and markets were opening up and, as with the subsequent privatization of the water industry, telecoms was seen as a sector requiring major new investment which would have represented an unwelcome call on Treasury funds. The Government's rhetoric of freeing up the sector from the shackles of public ownership struck a chord with the public and, in November 1984, the flotation of British Telecom (BT) was successfully completed. The sale of British Telecom was in two stages and raised £3.9 bn for the Government.

The next target, British Gas, was a rather different prospect. The provision of gas was seen much more as a basic need which could not be jeopardized. Whereas BT had a rather stolid, bureaucratic image – largely due to the long waiting list of people wanting to be connected to the system – British Gas had a much better image based on the efficient conversion of the supply system from town to natural gas in the 1970s and on the steady fall in the real price of gas that the development of the North Sea gas-fields had subsequently allowed.

From the time of the BT sale, some sections of the Conservative Party disliked merely converting public sector monopolies into private sector monopolies and wanted British Gas to be broken up into a competitive structure. By contrast, the government wanted to privatize British Gas quickly while also getting a reasonable revenue yield. In this debate, British Gas argued forcefully that splitting up the vertically integrated corporation would reduce the technical and economic security of public gas supply in Britain. British Gas was therefore privatized intact in December 1986 raising £5.6 bn.

The sale of the Trustee Savings Bank in 1986 should also be noted. This was not publicly owned, in fact its sale was based on the premise that nobody owned it and as a result, shares could be priced very low with no apparent loss to anyone. The profits made by the public from the subsequent sale of shares allocated in the flotation maintained the public expectation of regular windfall profits from company flotations.

1987–91

While the privatization of British Gas was controversial, the remaining public utilities, electricity, water, rail and postal services were, for differing reasons, even more contentious. In all cases, there was a general concern that a public service, previously available on reasonably equal terms would be lost, to the detriment of the weakest and poorest consumers. For water, the problem was the lack of any real prospect of competition and consumer choice being introduced. For rail, the inevitability that the rail industry would make large financial losses and that Government subsidies would have to go to private sector companies caused concern. For postal services, the risk that remoter parts of the country would lose the service was the issue and a successful campaign was mounted in 1993 opposing Government plans to privatize the Post Office. Public flotation of the water companies was completed in 1989 and rail privatization was underway in early 1996.

For electricity, the concerns surrounded the technical complexity of the supply system, the vital role electricity plays in modern life and electricity's unique characteristic of effectively not being storable. Various people then argued, and continued to argue right up to the share flotations, that privatization of the electricity supply industry would not be possible; but the Government was committed to its programme and confident of its ability to carry it out.

At the general election in 1987, electricity privatization was pledged in the victorious Conservative Party's manifesto but no details of the structure that would be chosen had been decided. At the Conservative Party conference following the 1987 election victory, the final shape of the industry had still not been decided, but the Energy Minister was then able to pledge that the Central Electricity Generating Board (CEGB) would not be privatized as a 'block monolith'.[5]

1991–95

By the time of the Conservative Party's re-election in 1991, little of the productive capacity of the British economy remained in public ownership. Facilities for which public ownership had previously been regarded as important on grounds of national security such as munitions factories, naval dockyards and research and development laboratories were now candidates for privatization. Only the coal, rail, postal services and the civil nuclear power sectors were obvious potential possibilities for the sale of major state-owned companies to the private sector. As noted previously, plans to sell the Post Office were abandoned and privatization of the rail industry is in progress at time of writing. The size of the coal industry had been much reduced following electricity privatization, and British Coal was privatized *en bloc* in 1994, in a trade sale rather than by public flotation, raising about £800 m for the Treasury. Partial privatization of Nuclear Electric and Scottish Nuclear is planned for 1996.

A new focus for the privatization programme had to be identified and this was found through a greater emphasis on the 'marketization' of services programme. Under this, service sectors such as education, health, garbage disposal and even parts of the civil service, for which total privatization was not a politically viable proposition, were exposed to market forces to a much greater extent with only those responsible for 'policy formulation' as opposed to implementation remaining in the public sector. For example, provision of hospital catering and supply of a school standards inspection service in state-owned facilities are now routinely allocated on the basis of competitive tender. The efforts to reduce dependence on a state-provided pension by encouraging people to set up their own pension funds should also be counted in with this programme.

Some of the secondary objectives of the earlier privatization programme were lost in this new phase: there was little revenue to the Treasury, no profits to the general public from share flotations and little opportunity to expand share-ownership; but the power of large trades unions was further eroded by the transfer of people from employment by the state to small, private-sector organizations. The ideology remained intact.

ELECTRICITY PRIVATIZATION

The size and complexity of the electricity supply industry meant that there could never have been any doubt that it would be more difficult than any previous privatization. The asset value of the UK electricity supply industry was estimated to be approximately 'four times as large as the total asset base of all the industries which were privatized in the first two Thatcher terms'.[6] Instead of dealing with just one company, as was the case with gas and telecommunications, the electricity supply industry for England and Wales was made up of 14 companies, 12 regionally based distribution companies, the CEGB and the Electricity Council.

The Scottish system was privatized at the same time, but to a different organizational model and this comprised a further two vertically integrated companies. The Northern Irish system comprised a single vertically integrated company. Its privatization was announced in 1988 and carried out in 1992 with a structure which was different again to either of those adopted in mainland Britain. However, the England and Wales system then included about 88 per cent of UK electricity consumption compared to 10 per cent for Scotland and 2 per cent for Northern Ireland and the model introduced for England and Wales is usually characterized as the UK model.

Regulation

When utility privatization was proposed, debates in Britain about the possible structure and regulation of private sector utilities were little developed outside the arguments of theoretical economists. From the start, the Government was clear about what it did not want. It did not want a regulatory system like the one that had evolved in the USA: this requires huge bureaucratic public bodies to administer it and dominates decision-making in the regulated utilities.

The US regulatory system is based on allowable rates-of-return on capital invested. Monopoly suppliers of utility services are allowed to make a rate of return, comparable to the rates earned in other sectors of the economy, on the agreed value of the assets employed in delivering the service (or rate-base). A risk with such a system is that, because the utility makes more money the greater the value of its assets, there will be an incentive to over-invest or to 'gold-plate' the system. This is countered in the US regulatory system by scrutiny of new investments to ensure that they are needed – the assets are 'used and useful' – and, to confirm that project costs were kept under control – the investment was prudently made.

In the electricity sector, these tests only came to prominence in the 1970s when electric utilities ordered far more generating capacity (often nuclear) than was warranted by demand growth and failed to ensure that project costs for these plants did not over-run significantly. Utility commissions, the state-level economic regulatory bodies, began to question these investments and to disallow some or all of the investment cost from the rate-base. This left shareholders, rather than consumers, to pick up the imprudent investment cost. The large sums of money at stake in the process of regulatory scrutiny of new investments meant that the proceedings were strongly contested and were often lengthy, controversial and expensive. They also meant that when utilities were trying to finance new investment, they had to convince financiers that the utility commission would look favourably on the investment. So the policy of regulators increasingly dominated utility decision-making.

The Austrian school of economists, through Michael Beesley and Stephen Littlechild, seemed to offer an alternative.[7] They proposed a system of incentive regulation whereby utilities are allowed to raise the charge for monopoly services by the general rate of inflation, adjusted by an efficiency factor. For example, for electricity, the transmission and distribution sectors are regarded as natural monopolies and the cost of this element of consumer bills is set by incentive regulation. Costs which can be determined by market forces are passed on in full to consumers. Electricity generation is assumed to be a competitive market and this cost element is set by the market. For most consumers, electricity supply was, at first, a monopoly service regulated by incentive regulation. As the restrictions on the ability of consumers to shop around for their supply contracts are progressively removed, the cost of this element will be increasingly set by the market.

In Britain, inflation is measured by the retail price index (RPI) and the incentive term was called the 'X' factor and the formula became known as 'RPI–X'. This formula has been applied throughout the utility privatization programme. The expectation was that 'X' would be a positive number and the cost of the utility service would therefore fall in real terms. This was expected to apply strong pressure on utilities to reduce costs but would leave the utility free to use its own judgement about the best investments to make to achieve these cost savings. This seemed to remove the need for regulators to make technical judgements. It was also proposed that the formula be set for several years forward so that businesses could have a stable and predictable economic environment in which to make long-term plans. This vision of infrequent and non-intrusive or 'light' regulation accorded with the Government's rhetoric on free markets and releasing entrepreneurial spirits. In practice, it is arguable that this regime did not differ so radically from the performance objectives covering controllable costs that were applied to nationalized industries in the 1980s (described in Chapter 2).

The other important aspect of regulation was the nature of the person or body that carried out the regulation. This broke with traditional British practice of using an impersonal committee structure which makes recommendations to a Government Minister which he or she can choose to accept or reject. For each utility, regulation is vested in a Director General for the industry with no direct and overt ministerial input to the process. The Director General is assisted in decision-making by a regulatory office so, for example, for telecoms there is an Office of Telecommunications or OFTEL. However, the decisions remain those of the Director General, in the case of electricity, the Director General of Electricity Supply (DGES).

The regulator must agree his or her decisions with the regulated body, but if agreement is not possible, the regulator must refer the issue to the government's Monopolies and Mergers Commission (MMC). This body would carry out a further investigation of the issues and make recommendations to the relevant Government Minister (for electricity and gas, currently of Trade and Industry) who makes the final judgement. In practice, an MMC investigation can be an almost open-ended process in which the MMC is free to conduct a very wide-ranging inquiry. This can be very disruptive to the operation of businesses and the outcome is difficult to predict. This risk of 'something worse' being recommended gives utilities a strong incentive not to challenge the Director General's decisions.

An additional economic regulator is the Director General of Fair Trading assisted by the Office of Fair Trading (OFT). This person has a broad responsibility to oversee the conduct of markets and make recommendations to the Trade and Industry Minister. The OFT has not yet been involved to any great extent in the privatized electricity supply industry.

The National Audit Office (NAO) has some influence where public money is involved. The NAO is the investigative arm of the House of Commons Public Accounts Select Committee which commissions the NAO to carry out investigations and which makes recommendations to Ministers on the basis of the NAO's reports. For the electricity supply industry, the NAO can only comment where public finance is involved which, in practice, means the conduct of the sale of shares and the nuclear power sector.

Environmental regulation has been through the various arms of the Department of Environment, notably, Her Majesty's Inspectorate of Pollution (HMIP) and through the Health and Safety Executive (HSE) which is independent of Government Ministries but reports primarily to the Secretary of State for Employment. In 1995, some of these functions for England and Wales were consolidated in the newly created Environment Agency. The MMC, OFT, NAO, HMIP and HSE all predated the privatization programme, although the extent and significance of their role have been changed by it.

THE CHOICE OF STRUCTURE

The government was also clear about what it wanted to avoid in the restructuring of the industry – a simple transfer of assets from public to private ownership with no new competitive mechanisms. However, it was far from clear, at least initially, about what structure it did want. The model for privatized network utilities has since become well-established: de-integration into the various component parts with natural monopolies operated by regulated companies and the potentially competitive parts opened to free competition. However, then, debates about the potential dangers of integration and the identification and separation of natural monopolies and the best way of introducing competition to areas where competition was possible were largely confined to theoretical economic forums. The British experiment was also the first time in the world that anyone had tried to design a competitive electricity supply system and so there was little precedent from overseas from which informed judgements might have been made as to what type of structure would prove workable in practice.

The choice of the new structure was far from being based solely on trying to identify the structure that would maximize economic efficiency. A number of other objectives had to be satisfied and a number of constraints taken account of. Some of the constraints were inevitable – the stock of power plants and the staff of the existing companies had to form the basis of the new industry. Some were cultural – since the Second World War, policies of both the major parties have tended to favour the adoption or retention of centralized structures. Some resulted from other political objectives that were being pursued, the most important of which were the general objectives of the privatization programme. Others resulted from experience with earlier privatizations.

The option that, after careful consideration, retaining the industry in public ownership was the best choice was therefore not one that was available to the Secretaries of State for Energy charged by Mrs Thatcher with carrying through the process. Decisions that could be represented as policy 'U'-turns had become anathema to the Conservative Government, and the Treasury was counting on the revenues. In addition, the re-organization would have to be completed within the time-span of one Parliament. Any risk that the process would not be completed at the time of a general election could not be contemplated on grounds of economic and political risk. The revenues accruing could also be used to party political advantage by cutting taxes just prior to a general election. Although a British Parliament can run for up to five years, the period between elections (decided by the Government) is usually expected to be no more than four years. Allowing for the period leading up to an election when little of substance can be accomplished, this meant that the whole process, from the start of design of the new structure to completion of the flotation of the companies, was crammed into a period of less than four years.

The crucial period for putting the new structure in place and completing all the various stages of the legislative process, devising the basis of regulation, and setting up the licensing system for the new electricity companies was only two years – March 1988 to March 1990. This politically determined timetable allowed no delay and no change of mind. Any proposal to back-track would have received short shrift from the architect of the privatization process, the Prime Minister. The Government's majority in the House of Commons was large and this gave a sense that anything could be achieved given the will.

Objectives and Constraints

All the objectives that had been pursued with the overall privatization programme were important to electricity privatization. These included: generation of revenue for the Treasury, and at the time of electricity privatization the Treasury had a target of proceeds of £5bn per annum from sales of assets; a sale of shares to the general public which would either expand share ownership or give quick, assured profits; and a reduction in the power of trades unions. A particular constraint on the process was a commitment contained in the Conservative Party's May 1987 election manifesto to promote nuclear power development,[8] a commitment which many predicted would be difficult to reconcile with privatization.[9]

The objective of reducing trades union power was particularly important with the National Union of Mineworkers (NUM), the main coal-mining union, the prime target. As in the early 1970s when Edward Heath was Prime Minister, Mrs Thatcher and her Governments from 1979 onwards recognized the power of the NUM to jeopardize coal supplies to power stations. This power to 'turn the lights out' was seen as a danger to the Government and its policies, a threat they were determined to remove.

The electricity market was the only significant market for British coal, and British coal accounted for nearly 80 per cent of British power generation. Changing the dynamics of the electricity market represented a good opportunity to reduce NUM power. Whilst this was not admitted at the time, subsequent memoirs from the cabinet ministers involved have confirmed this motive.[10] The wish to break the power of the NUM could not be stated in those terms, but the Government used code words such as 'diversification' which few failed to decode.[11] The protected market for British coal could not be expected to be sustained in a privately owned system and private companies would not be so vulnerable to the political pressures that weakened the electricity supply industry's negotiating position against the British coal industry.

The objective of providing the general public with an opportunity to make easy and assured profits was more difficult to arrange if the option of privatization without restructuring was ruled out. Profits could be easily guaranteed for utility sales where the central monopolies were only marginally eroded and recognisable companies left intact, but the introduction of competition and of new companies with no track record is inevitably accompanied by an increase in economic risk. So competition had to be muted, at least for the first few years, to ensure that the companies did not fail and the public lose money. Dramatic changes to the structure that would make the final shape of the industry less predictable would also reduce the proceeds from privatization and would, on those grounds, be unappealing to the Treasury. Privatization had to be completed before the next general election and, in practice, this meant that it could not be planned to take much longer than 3–4 years. Creating a large number of new companies would have been time-consuming and so a minimum of restructuring was essential.

The risk of early failure of the privatized companies was further reduced by opening up markets to competition only gradually. This was achieved by a series of transitional contracts imposed by the Energy Minister that effectively meant that, at least for the first three years, the

Power Pool was no more than a shadow market and that most final consumers had no choice of supplier. At the head of this process were contracts between British Coal and the generation companies. While the Conservative Party was determined to break the power of the NUM, a full opening up of the generation market might have seen a chaotic collapse in the market for British coal in favour of imported coal. This would have been politically dangerous and was therefore not acceptable. The three-year contracts specified volumes of coal not far short of those taken by the CEGB, but at prices that fell significantly in annual increments in real terms. This fall in coal prices served the dual purpose of apparently giving the British coal industry a sporting chance, albeit under intense pressure, and it also guaranteed that there would be economic rent which could be distributed to consumers, shareholders, staff or the directors of the companies (for further discussion of the impact on the coal industry, see Chapter 6).

The next layer of contracts was between generators and distributors. These so-called contracts for differences were set at prices that had little or no relation to Pool prices. In simple terms, for example, if the Pool price was 2p/kWh and the contract price was 3p/kWh, the generator would receive 2p/kWh from the Pool and the distributor would purchase power at 2p/kWh from the Pool, but the contract for differences would specify that the distributor would reimburse the generator the 1p/kWh difference between contract and Pool price. These contracts protected the generators from competition from new generators that might be able to buy fuel more cheaply and meant that little change in the market share of the generators was expected to be possible for the duration of these contracts. The final layer of protection for investors was the retention by the new distribution companies of much of their geographical franchise for the first eight years of operation of the new system.

While these transitional contracts secured the short-term future of the British coal industry, the new structure inevitably led to the loss of NUM power. This power had derived from the effective monopoly British coal had for coal-fired power generation. During the 1980s, this resulted in coal being supplied at prices which were sometimes significantly above those of the emerging international coal market. In these circumstances, any disruption to supplies of British coal would have immediate implications for the security of supply. While a case might be made to retain this privileged position for British coal on grounds such as independence from unpredictable world fuel prices and benefits to the balance of trade, these national strategic arguments would not impress a privately owned company competing hard for its market share. If British coal was to retain its market in the new system, it would have to match world prices and offer secure supplies. These conditions would give the mining unions little scope for taking effective industrial action which did not severely damage the long-term interests of the coal industry.

The Government commitment to support the development of civil nuclear power in the private sector was, in part, a continuation of the support that UK Governments of all complexions had given to nuclear power. Mrs Thatcher in particular was a strong supporter of nuclear power and a visit to France early in her first term of office left her with admiration for the French Pressurized Water Reactor (PWR) programme and also the fast-breeder reactor. An additional explanation to this general political support was that it represented the paying off a 'debt' to the British nuclear industry resulting from the miners' strike of 1984–85. Then, the CEGB managed to squeeze extra power out of the nuclear power stations and this reduced the rate at which coal stocks were used up, contributing to the defeat of the strike.[12]

While privatization could be seen as an instrument for breaking the power of the NUM, privatization was not expected to help promote nuclear power. It was understood that factors such as its high capital intensity, the heavy technological demands it places on the owner and the financial risks, might make privatized companies reluctant to invest in new nuclear power plants. To counter this, the Government proposed that the nuclear power plants be owned by a large generating company, which would inherit two-thirds of the CEGB's capacity, and this would give the technological and economic strength to 'shelter' the nuclear plants. In order that there be an effective counterweight to this large company, all the other capacity was to be placed in one company. The distribution companies would also be required to purchase a proportion of their power, to be set by the Government, from nuclear plants so that when old nuclear plants were retired, they would have to ensure that new plants were built.

THE NEW STRUCTURE

The Government's February 1988 White Paper on privatization finally gave some details of the structure first proposed.[13] The distribution companies, renamed Regional Electricity Companies (RECs), would be sold intact, but the CEGB would be divided into three parts. As discussed previously, the structure for the generation sector was tailored so that the nuclear power plant could be accommodated in the private sector and involved the creation of only two generating companies, National Power, the larger company with the nuclear power plants, and PowerGen. The non-nuclear power stations were shared out between the two companies very carefully on the basis of the fuels used and the location so that neither of the two generators had a significant advantage in its generation cost base.

Details of how the system would run in practice were still sketchy at this point. A Power Pool was proposed in order that the basic philosophy behind electricity privatization be met, that all areas of electricity supply that are not natural monopolies would be opened to full and vigorous competition as soon as was practical. The Power Pool was to be the main price-setting arena into which buyers and sellers could place bids. Contracts of limited term outside the Pool were anticipated for power purchasers that needed greater predictability in their costs, but pricing of these contracts would tend to use Pool Prices as their bench mark. These mechanisms were expected to be competitive enough that no need for routine regulation of the generation sector was anticipated.

It was also expected that the supply of power to final consumers could be made fully competitive. Clearly there was no question of duplicating the actual distribution infrastructure, but if the network was a resource open to all, on non-discriminatory terms, competition could be introduced. For example, if a supply company could purchase power more cheaply, or it could supply customers incurring lower overheads, it would be able to offer cheaper terms to consumers. From Vesting Day, 1 April 1990 (the day the new structure came into operation), consumers with demand greater than 1 MW were able to negotiate their supply from any licensed supplier. This limit was reduced to 100 kW in April 1994 and is scheduled to be removed altogether in April 1998. Customers negotiating the supply of power on individual contracts are assumed to have open to them a competitive enough market that regulation is not required. However, as a back-up for those not wishing to exercise this right, the local REC is currently still obliged to offer supply to all consumers with demand less than 10 MW under the terms of a published and regulated tariff.

The remaining area, which amounts to the physical infrastructure of the network, the high voltage transmission system and the local distribution system, would be regulated by OFFER, using incentive regulation.

PROBLEMS ON THE WAY

The period of privatization was marked by false starts, leaks of plans and failure to meet target dates. In part, this reflected the novelty of the process to all those involved, outside as well as inside Government. The private sector, with little experience of analysing this sector, was no more impressive than the civil servants and investment analysts did little to alert the Government to problems in its approach. The most crushing indictment of the lack of understanding of the sector by civil servants and financial analysts was the failure to alert the Government to the total impracticality of the plans to privatize nuclear power.

The Magnox plants had very little operating life left and massive liabilities for spent fuel disposal and decommissioning, and the advanced gas-cooled reactors (AGRs) then showed little sign of becoming reliable generators of power. For future plants, the higher rate of return required by private markets (now estimated to be more than 12 per cent real per annum) compared to that required by Government (during privatization raised from 5 per cent to 8 per cent) seemed likely to make new nuclear power stations uncompetitive. Far from being the cheapest generation option as the CEGB had long argued, using private-sector investment appraisal methods and required rates of return, it turned out that power from the Sizewell B PWR, on which construction had just started, would be approximately twice as expensive as the alternatives. These factors were obvious to almost all experienced energy analysts. In April 1987, before the Conservative Party was re-elected, Holmes et al wrote, 'It seems the only possible route is to leave the nuclear industry in state hands'.[14] In July 1988, the House of Commons Select Committee wrote:

> The independent witnesses we examined were unanimous in their view that ... private generating companies would be most unlikely to build new nuclear power plants.[15]

There was some acknowledgement that the high capital costs and the commercial risks of nuclear power might mean that generating companies would not opt for nuclear power. A non-fossil fuel obligation (NFFO), expected to be about 20 per cent, was therefore to be placed on distribution companies which would require them to obtain a certain percentage, to be specified by the relevant Minister, of their power from non-fossil fuel sources. If distribution companies could see they would not be able to meet this target in the future, perhaps due to plant retirement or demand growth, they would be obliged to contract someone to build additional non-fossil capacity.

It soon became clear that no such long-term obligations could be placed on private sector companies and the NFFO quickly became no more than a mechanism to ensure that the existing nuclear power plants were used to their full extent. This change in policy was particularly embarrassing to the CEGB which, at that time, was arguing the case for a new PWR at a public inquiry on the basis that the plant would be needed to fulfil the expected level of the NFFO.[16]

Why it took two years for the Government to act on these and many other similar warnings that nuclear power could not be sold is not clear. The CEGB itself repeatedly warned the Department of Energy that electricity privatization would be jeopardized if Government financial guarantees were not in place to cover the long-term nuclear risks and liabilities. In part, it must be due to the naiveté of the Department of Energy and the City, but it may also have been due to the unwillingness of Ministers to listen, particularly when the message would have been so unpalatable to Mrs Thatcher. The warnings may have been interpreted as ideologically motivated statements by parties with an interest in obstructing the process of privatization. Even when the facts could be ignored no longer, the Government retreated slowly. In July 1989, the then Secretary of State for Energy and the architect of the privatization proposals, Cecil (now Lord) Parkinson, in practically his last act in that role, withdrew the Magnoxes from the sale.

By then, the process of privatization seemed close to becoming so delayed that it might have to be abandoned until after the next election. Parkinson was replaced by John (now Lord) Wakeham. This move was seen by much of the media as the replacement of a weak minister by a 'political fixer'. The emphasis switched from trying to create the most competitive market solution, to simply completing the job in the time available. The commitment to nuclear power could not immediately be abandoned, but the remainder of the nuclear power sector had to be withdrawn from the sale in November 1989. However, the demands of the timetable were such that no more than a minimum of re-thinking for the remainder of the generating sector was possible. No structural changes were introduced, except that National Power was to be privatized without the nuclear power plant. This was now to be held by a new, publicly owned company, Nuclear Electric. Wakeham has since claimed that had it been known from the outset that nuclear power could not be privatized in the sort of competitive structure proposed, it would have been preferable to split up the generation business into much smaller units.[17]

Nuclear Electric was to be compensated for the high cost of nuclear power from the proceeds of a fossil fuel levy (FFL) of about 10 per cent charged on all sales of power. The FFL has yielded £1.2–1.4 bn per annum and has provided 40–50 per cent of Nuclear Electric's income. Privileges under the NFFO and the existing nuclear operating regime (ENOR) effectively meant that nuclear power plants would be used whenever available. However, a moratorium was placed on all new orders for nuclear power plant, excluding the Sizewell B PWR, which was then already under construction, until at least 1994, when a Government review of nuclear power policy was to be completed. The European Commission forced one major concession on these proposals, that the FFL be phased out by 1998.

Another area causing problems was the Power Pool. This was to be the centrepiece of the system, and would be an open, transparent spot market for power. Here, all generators, large and small, would be able to compete on equal terms to sell their power, and power purchasers would also be able to put in their bids to buy power. In autumn 1989, the proposed new double Pool structure had to be abandoned because the complex computer software needed for its operation could not be completed in time. It was replaced by a new system, which necessarily was much simpler so that it could be in place by 1 April 1990, Vesting Day, when the new companies came into formal existence and the Power Pool began operation. This simplified Pool, based on existing software used by the CEGB for the dispatch of plant, removed the possibility of buyers placing bids for the price at which they were prepared to buy power, and made it a market-clearing mechanism for generators.

In December 1990, the RECs were sold to the public in a share issue that was heavily oversubscribed.[18] The shares, sold at £1, rose immediately to about £1.50, effectively generating instant windfall profits for those purchasing shares. The transmission sector was separated off into a new company, National Grid Company (NGC).[19] Ownership of NGC was initially passed to the 12 RECs. The risk that ownership by the RECs of the company controlling the grid, particularly the despatching system, would lead to grid policies which would unfairly favour the RECs was countered by provisions which restricted the extent to which RECs can influence the policies of the NGC. In December 1995, at the instigation of the regulator, the RECs were required to sell nearly all their shares in the NGC on the stock market.

In March 1991, 60 per cent of the shares in the two generating companies were sold for £1 each, again in a heavily oversubscribed sale, and prices rose immediately to £1.37 in early trading.[20] The other 40 per cent of shares remained in UK Government hands until they were sold early in 1995. The ownership of the nuclear plant is under review and the potentially saleable part of Nuclear Electric, the AGR stations and Sizewell B PWR, has been combined with Scottish Nuclear's AGRs in a new company, British Energy, which the Government proposes to privatize in 1996. The Magnox stations are now owned by Magnox Electric, which will remain in public hands and which, it is proposed, will become part of the state-owned fuel cycle and waste-treatment company, British Nuclear Fuels Ltd (BNFL).

Scotland and Northern Ireland

Scotland

A very different, less radical structure was chosen for the concurrent privatization of the Scottish system. Previously, Scotland had been supplied by two interconnected systems which were fully vertically integrated and were synchronized to the England and Wales system. The South of Scotland Electricity Board (SSEB) was much the larger system with nearly 4 million customers and about 6 GW of plant. About 60 per cent of its power came from its two AGR nuclear power stations, the newly completed Torness and Hunterston B, with a third nuclear plant, the Hunterston A Magnox station being closed in 1989. Most of the rest of its plant was coal-fired and had low utilization.

By contrast, the North of Scotland Hydro-Electric Board (NSHEB) had only about 1 million customers and 3 GW of plant, split between hydro and fossil fuel plants. NSHEB was rather a different entity to the other electricity companies having a proud history since the 1930s of participation in the economic development of the Highlands through exploitation of the hydraulic resources.[21]

Had the proposals to privatize the nuclear power sector in England and Wales not broken down, it is likely that the Scottish system would have been privatized intact with little change to the method of operation. The problems revealed in privatizing nuclear power in England and Wales also led to the withdrawal of the nuclear plant from sale in Scotland. These were retained in the public sector in a new company, Scottish Nuclear (SN), and the integrated generation, transmission, distribution and supply companies, SSEB and NSHEB, were privatized as Scottish Power (SP) and Scottish Hydro-Electric (SHE) respectively.

There appear to have been a number of practical factors behind the decision not to attempt to create a competitive market on the English model. First, the Scottish system is much smaller than the England and Wales system and it would have been difficult to create competing generation companies; second, the dominance of the nuclear power plants, which supply more than half Scotland's power and which must, for technical reasons, be given priority in dispatching leaves little realistic scope for other generators; third, the lack of pre-existing distribution companies in Scotland meant that new companies would have to be created if full competition in supply was to be introduced. From a political point of view, a large majority of the Scottish electorate had consistently voted against the Conservative party and was unsympathetic to the Government's objectives. While the companies were privatized, the decision not to break them up can be seen as a minor concession to Scottish sensibilities.

The dominance of nuclear power in the generation mix meant that a levy such as was imposed in England was not appropriate. Nuclear power was guaranteed to be used whenever possible and it was purchased by SP and SHE at a guaranteed premium price which was calculated to be sufficient to cover SN's costs. Like the FFL, the premium over Pool Prices in England and Wales was to be phased out by 1998. In practical terms, the absence of any operating Magnox plants and the fact that the AGRs had performed better than the equivalent plants in England and Wales meant that the implicit subsidy was not as large as was required in England and Wales. It was anticipated that the substantial surplus of plant, partly created by the completion of Torness, would be used to increase exports of power from Scotland to the England and Wales system, although the capacity of the links limited the potential for power exports.

Regulation of the transmission, distribution and supply elements of the business are via the RPI-X formula under the DGES with the assistance of a Scottish office of OFFER. Generation is not systematically and formally regulated.

The new system was brought into operation on the same day as that for England and Wales, in 1990, and shares in the new companies were sold in June 1991 at £1 per share. The premium on early trading was less than in the earlier flotations but was still about 20p per share.[22]

NORTHERN IRELAND

The new system for Northern Ireland represented a third model, significantly different from those adopted for England and Wales and for Scotland. The absence of any nuclear power plants, which had distorted the process in England and Wales and had limited the options in Scotland gave the Government a much freer hand to design a system that met their objectives. The long-term potential for competition contained in the new structure adopted makes the system of particular interest.

Previously, Northern Ireland had operated under a similar arrangement to Scotland but with a single generation, transmission and supply company, Northern Ireland Electric (NIE). While NIE was comparable in size to NSHEB, there were other factors which influenced the structure chosen and meant that a 'Scottish' solution was not adopted. The Irish electricity supply system was totally isolated, with no link to Britain and the 300 MW link to Southern Ireland temporarily abandoned because of terrorist action. This and the lack of an indigenous coal industry meant that the electricity system was high-cost. For political reasons, the British Government had given a subsidy such that the price of electricity in Northern Ireland was no higher than the average for mainland Britain. This subsidy was particularly important given the absence of gas from Northern Ireland. The old manufactured gas ('town' gas) system had fallen into disuse and no connection existed by which to bring natural gas in from Britain. This meant that domestic space heating was inevitably more expensive than in most of Britain, a disparity that would have been even more pronounced without the subsidy. In addition, privatization was carried out three years later than in mainland Britain and would be expected to benefit from experience there. The White Paper on the privatization was published in March 1991, a year after completion of the new structure in England and Wales.[23]

Three other energy policy objectives were rolled into the privatization process: the laying of an electricity cable to Scotland to allow trade between the systems and to improve the security of supply in Northern Ireland; the building of a natural gas pipeline from Scotland to Northern Ireland so that the economic and environmental benefits of access to this fuel would be available to Northern Ireland; and the removal of the electricity subsidy.

The new structure for the industry was as follows.[24] The four large power stations were sold separately in June 1992, via a trade sale, to three different companies. A large modern coal-fired station (Kilroot) and a very old oil-fired station (Belfast West) were sold to Nigen, a consortium based on the Belgian company, Tractebel and the US company AES. A large oil-fired station (Ballylumford) was sold to a subsidiary of British Gas, Premier Energy, on the understanding that British Gas would build the gas interconnector and convert the station to gas-firing. A small oil-fired station (Coolkeeragh) was sold to its employees. Kilroot and Ballylumford are expected to provide the mainstay of the system for at least the next decade. The remainder of the NIE including the transmission, distribution and supply businesses was sold by flotation in 1993. The shares were priced at £2.20 and the offer was more than four times over-subscribed, although the price rose only by 26p in early trading.

Regulation is carried out by a new regulator, the DGES for Northern Ireland. The new generation companies were protected by long-term contracts for sale of power of up to nearly 30 years (in the case of Kilroot) although the regulator is now examining the possibility of introducing a Power Pool, modelled on that used in England and Wales. How this could be made compatible with the long-term generation contracts or work economically given that there will be only three power stations in a few years is not clear.[25]

CONCLUSIONS

The British policy of privatization has been taken up worldwide as a major tool of economic policy. In some cases, privatization has been applied to companies operating in global commercial markets and here the judgement is, in principle, a simple one of whether the claimed benefits of the greater efficiency of private companies more than outweigh the loss of national strategic control that privatization brings with it. The privatization of utilities is a more complex affair, often involving three separate, not necessarily connected elements. The first is clearly a shift in ownership from the state to private hands. The second is a corporate restructuring leading to the creation of new or radically re-shaped companies. The third is a change in procedures by which the sector operates, usually involving an injection of competitive procedures.

Previously in Britain, utility privatizations involved only the first of these elements and electricity privatization in Scotland also involved little more than a change of ownership. The privatization of the electricity supply industry of England and Wales and, to a lesser extent, that of

Northern Ireland, which involved all three elements must be seen as path-breaking experiments by Government in industrial economics.

The simultaneous execution of all three elements placed inevitable constraints on the process. The objective of selling the companies to private investors, especially if a public flotation was chosen, was incompatible with a system that put the companies at serious commercial risk for at least the first 5–10 years. This meant that restructuring of the sector had to be limited. Creating a large number of new companies with no track record would have also adversely affected the price obtained for the assets. Nevertheless, for England and Wales, a number of important new and untested procedures were put in place, such as the Power Pool. The impact of these new procedures was heavily limited by government-imposed transitional arrangements, such as coal supply contracts, which have proved to offer a very favourable environment for the new companies.

There is little evidence, however, that the government set out with the intention of designing a radical new system, nor is there any evidence that there was any vision of what the endpoint of the reforms would look like. The system evolved in response to criticisms of earlier utility privatizations and the structure chosen was heavily shaped by other government objectives, such as reducing the power of the coal-mining union.

Two other major aspects of electricity privatization are important. First, it included all parts of the industry even when it was unclear whether the intended structure would work or where there was no logic behind the structure. For example, allowing the RECs to own the two natural monopoly elements of the industry, the high voltage transmission and the local distribution systems, while also keeping the increasingly competitive supply business is hard to justify. Second, the timetable was politically driven and could tolerate no significant change of plan.

It should not be surprising therefore that this rather unstructured and confused process has resulted in a system which, at best, will take some time to settle down and, at worst, may prove to have very serious fundamental defects in the longer term.

Part II

THE RECORD, 1990–96

4

The Development of Competition

Steve Thomas

Introduction

What marked electricity privatization out from previous utility privatizations was that the Government attempted to engineer competition rather than simply waiting for it to emerge by some natural process. They did this by restructuring the industry and by imposing competitive mechanisms. This chapter examines how successful these measures have been so far in encouraging competition.

The section on restructuring identifies the consequences for competition of the original structure chosen. It then looks at structural developments since privatization, distinguishing structural change instigated by the regulator and structural change inspired by the market. It then examines the potential consequences of the proposed privatization of some of the nuclear power plants.

The chapter then looks at the development of competition in generation, examining how well the Power Pool has met the objectives that were set for it. It reviews the impact of the 'contracts for differences' which now dominate power purchasing and the regulatory attempts to reduce the power to set the Pool price of the two large, privatized generating companies. It then examines the security of supply of the system. Finally, the introduction of competition to end-consumers is assessed, particularly the progress that has been made towards the proposed opening up to competition in 1998 of the market for domestic consumers.

EFFECTS OF THE INITIAL STRUCTURE

The key factors in determining the new structure were to identify the areas where competitive forces could apply and those which were natural monopolies. While the Government did claim that regulation would wither away as competition took over, this claim is hard to take seriously for the transmission and distribution sectors. It is inconceivable that a privately owned company would be allowed to operate a monopoly service without some form of explicit regulation. It may be that the Government believed that scientific advances would bring new technologies which would mean that transmission and distribution were not monopolies, as has happened to a certain extent with telecommunications, but no such advances are currently expected. The potentially competitive areas are therefore generation and supply.

As discussed in the previous chapter, the generation side was restructured with the objective of setting up competing companies. However, the Government's objectives for nuclear power and perhaps the logistical difficulties of creating several new generation companies from the Central Electricity Generating Board (CEGB) meant that only two private sector generation companies were set up initially. The Government predicted that new entrants to the generation market would quickly emerge to produce a wider field of competitors. How accurate this prediction has turned out to be is discussed below.

No restructuring was undertaken to encourage the emergence of a supply market and the incumbent distributors initially retained most of their monopoly privileges over their franchise territories. However, a timetable was set for the progressive removal of the franchise. Customers having a maximum demand of more than 1 MW were able to choose suppliers from Vesting Day,[1] customers with a maximum demand of 100 kW were able to choose from April 1994, and it is planned that all remaining customers will be offered choice from April 1998. The non-franchise customers are able to choose from any supplier licensed by the regulator and this immediately brought in the main generation companies as well as any of the 12 Regional Electricity Companies (RECs). How this market has developed is also discussed below.

For the natural monopolies, the high-voltage national transmission grid and the low-voltage local distribution infrastructure, there was thought to be no merit in breaking down geographically defined monopolies. Effectively this would have led to a wasteful duplication of the

Table 4.1 Structure of the supply market

To market	*Number of customers*	*Percentage of total demand*	*Percentage of market to switch by year 1*	*Percentage of market to switch by year 5*
>1 MW	5000	30	31	61
0.1–1 MW	45000	20	ca 50	n/a
<100 kW	22 m	50	n/a	n/a

Source: L Phillips (1995) *European Electricity Liberalisation: Lessons from the UK*, MC Securities, London, pp 6–8.

network and might have led to the loss of the benefits of scale that a fully interconnected system brings. Regarding the objective of encouraging competition, the priority with these sectors was to ensure that access was available on non-discriminatory terms which fully reflected the costs incurred. An important step to achieving this was to compel owners of the network to derive and publish standard, non-discriminatory tariffs for the use of the network. This was done, although how far the tariffs reflect the costs incurred, particularly for the transmission network, is an issue we return to later.

The other major issue was ownership of the networks. A possible risk was that if companies that owned the network operated in the competitive parts of the industry, they would be able, if not by discriminatory access and prices, then by information advantage, to favour their competitive business. This risk can be combated by prohibiting companies that own the network from operating in the competitive parts of the business – so-called de-integration.

De-integration was only partly applied in England and Wales. The high-voltage grid which was previously owned and operated by the CEGB was transferred to a new company, the National Grid Company (NGC), but this was owned by the 12 RECs. To counter the risk that the RECs would manipulate the policies of the NGC to serve their own objectives, strict restrictions were placed on the extent to which they could influence the policies of the NGC.

The NGC was also given a small presence in the generation market through ownership of the two hydro-electric pumped storage plants which are sited in North Wales. While these comprise less than 2 GW of capacity, they are of key importance to the operation of the system. The rationale for pumped storage plants is that they consume electricity by pumping water from a low to a high reservoir when electricity is cheap and generate electricity by allowing the water to fall back through electricity generators when the price is high. This type of plant has two main advantages: it effectively stores electricity and its rapid response time – a few seconds from zero to full power – means that sharp peaks in demand can be met cheaply. This key role in assuring system stability was originally seen as being best vested with NGC.

While there was some attempt to ensure that there was no potential conflict of interest for the owners of the high-voltage grid, there was much less stress placed on restructuring the low-voltage distribution side of the business. The pre-existing distribution companies were privatized intact, although they were required to make an accounting separation between the distribution and supply sides of their business. As noted previously, they were given ownership of NGC. They were also allowed to own generating plant, with the proviso that they should not be contracted to these plants for more than 15 per cent of their power purchases. Many of the RECs, in the so-called dash-for-gas, quickly came close to filling this quota and, had market conditions and other strategic objectives favoured the installation of further gas-fired capacity, there would have been pressure on the regulator to lift this quota.

Two issues that were not addressed in the initial restructuring were policy on mergers, and policy towards the nature of the ownership of the companies and the extent to which the companies could diversify out of the UK and also into other sectors in the UK. The concern on mergers focused especially on the RECs – a re-merger of PowerGen and National Power without nationalization was not politically or commercially plausible. However, the prospect of RECs wishing to merge was seen as very real and carried with it the risk that the supply market would be too narrow to be genuinely competitive. The Government retained some right of veto over the ownership of the RECs through a 'golden share', but this was allowed to lapse in March 1995 and, as widely expected, was followed by a number of takeover bids. The Government retains 'golden' shares in NGC and the two large privatized generating companies. It used the latter in May 1996 to discourage the US utility, Southern Company, from mounting a take-over bid for National Power.

In part, the general issue of ownership raises an emotional fear that if the companies become part of a diversified, perhaps overseas-owned, company, the ability for Government to control a strategically important resource would be reduced or even lost. Since the Government's policy was, in part, based on an assumption that electricity could be treated much more like any other commodity than previously, this was not a factor in determining the structure. A more immediate risk to the attainment of a competitive market was that companies owning the monopoly elements might, undetected by the regulator, be able to transfer profits out of that part of their business into unregulated businesses. This would protect the unregulated business from competitive threat and would result in customers having to pay excessive charges for the use of the system. These issues are taken up in Chapters 10 and 11.

The new electricity system for England and Wales is frequently characterized as having been restructured and de-integrated, but this observation does not stand up to scrutiny. In the previous chapter it was argued that creating an economically optimal structure was far from the only criterion when the new structure was chosen. Other factors such as the need to protect nuclear power, to create financially secure companies both to attract potential shareholders and to complete the changes within a tight time-scale were at least as important. As we saw in Chapter 3, in 1989–90, the Government's overriding objective was to secure a successful flotation within a tight timetable, influenced by electoral considerations.

The main act of de-integration was the creation of a new company which would own and operate the high-voltage grid. Since ownership of this company was placed with the distribution companies and the new company had generation interests, the new system was scarcely less integrated than previously in that respect. The main mechanisms concerned accounting and control with the separation of the distribution and supply businesses of the RECs, the generation and transmission interests of NGC and the restrictions on the degree to which the RECs could influence NGC. But, in a number of respects, the system became more fully integrated: the generation companies were able to enter the supply business to a significant extent and the RECs were able to integrate back into generation.

For the competitive sectors of the market, some restructuring was carried out in the generation sector with the splitting of the CEGB into two private-sector generation companies and a publicly owned nuclear company. The latter was not an intended result and, because the company is subsidized and effectively does not compete in the Pool or for power contracts, its existence was not expected to bring any competitive benefits. A market with two main suppliers which have detailed knowledge of each other's plant and generating costs is clearly not a satisfactory basis for competition. The Government's stated belief that new entrants would come into the market and quickly remove any duopoly power was therefore crucial for the structure adopted.

The decision not to split the RECs into wholly separate distribution and supply companies may have been made for several reasons. One may have been that, although the Area Boards were privatized intact and with similar business scope to previously to form the RECs, the market orientation of the system made this role far more demanding. Previously they had bought power at a standard tariff from the CEGB and supplied a captive market. Their technical credentials for owning and maintaining a distribution network were therefore strong, but their commercial credentials were negligible. Since the determinant of success in the supply business is likely to be commercial rather than technological expertise, the Government would essentially have had to create 12 new supply companies to retail electricity. This would have been a time-consuming and expensive task and would have created companies with no track record and a flotation would have been difficult to market. Since few consumers had any initial choice of supplier, the decision not to split distribution and supply immediately may have been made on the basis of waiting until the market was fully open before deciding whether structural changes were necessary.

CHANGES IN STRUCTURE SINCE VESTING DAY

The structure of the industry has altered significantly since privatization, sometimes at the instigation of the regulator and sometimes as a result of strategic decisions by the market players.

REGULATORY-INSPIRED CHANGES

The regulator has tried to correct imperfections in the privatized structure. He regarded the RECs' joint ownership of NGC as potentially unhealthy, although there was no suggestion that the RECs had actually exploited their position. In December 1995, the RECs were required to sell their holding in NGC via the stock market, where NGC shares are now traded. The regulator also saw NGC's ownership of the pumped storage generating plant as creating a potential conflict of interest, although again, there was no suggestion that NGC had unfairly exploited its market position. The pumped storage plant has proved to be as strategically important as anticipated: for example, the Pool price was set by the pumped storage plant for 15 per cent of the time in 1994.[2] The regulator therefore required NGC to divest its generation business, known as First Hydro. National Power and PowerGen were explicitly excluded from bidding for it. These plants were sold by trade sale to an American company, Mission Energy, in January 1996 for £680 m.

On the generation side, the regulator has become concerned at the continued ability of National Power and PowerGen to set the Pool price. Despite the completion of more than 5000 MW of gas-fired plant owned by companies other than National Power and PowerGen and the presence of a significant volume of imports of power from France, National Power and PowerGen set the Pool price for 85 per cent of the half-hour periods in 1994 (the pumped storage plants setting the price for the remainder). While the market shares of National Power and PowerGen have been eroded, their market power has therefore remained almost fully intact. To counter this, the regulator obtained their agreement in 1994 that they would use their best efforts to sell 6000 MW of their coal-fired plant to new competitors by the end of March 1996. His bargaining position in obtaining this agreement was the threat of a referral of this duopoly to the Monopolies and Mergers Commission (MMC). Such a referral could involve a long, disruptive inquiry with an unpredictable outcome. For example, the MMC might recommend a radical break-up of the generation side, which would be disastrous as far as National Power and PowerGen were concerned.

In the autumn of 1995, PowerGen agreed to sell its 2000 MW of plant to an REC, Eastern Electricity, which was taken over by the Hanson group earlier in the year. In April 1996, National Power also agreed to sell its 4000 MW quota to Hanson. The Government decision in April 1996 not to allow takeovers of RECs by National Power and PowerGen delayed completion of these sales. This divestment would contribute to some extent to the reduction of the duopoly hold on the market. However, Eastern already has a substantial generation business and, if this sale is completed, it will reinforce the trend towards an integration of generation and supply (this is examined in more detail in Chapter 10).

MARKET-INSPIRED CHANGES

Of the market-generated structural changes, the most important is the emergence of new generation companies or independent power producers (IPPs). About 7000 MW of new generating capacity larger than 100 MW has been ordered by 12 new companies. All the plant is of the combined-cycle gas turbine (CCGT) type and burns exclusively gas. This appears a vindication of the Government's prediction that new generation companies would quickly emerge to provide additional competition for National Power and PowerGen, but closer examination of their ownership and their method of operation suggests the competitive gain has been rather small.

Nearly all the capacity has been built by companies in which the major stakeholder is one of the RECs, and the justification for their construction appears to have been more strategic than economic. This capacity reduced the dependence of the RECs on what they saw as the duopoly power of National Power and PowerGen and it expanded their business into an area where their profits were not regulated.

From a financial viewpoint, the plant seemed to be a safe investment because contractual and insurance cover was available to cover all the obvious risks. The gas for most of these plants was bought from British Gas on 15-year supply contracts at prices which were then seen as low and which were closely indexed to general inflation rather than to the oil price. Gas supply was not dependent on the output of a particular field and, while the gas contracts were 'interruptible', the periods of interruption and their consequences gave little cause for concern. The entire power output was sold under 15-year contracts, to match the gas contracts, to a number of RECs with the plant owner as the main customer. These power contracts allowed full pass-through of the gas purchase price.

These plants were planned against a strong expectation that Pool prices and the contract price of power from National Power and PowerGen would rise significantly once the initial contracts imposed by the Government had expired. If the cost of the output of these plants was forecast to be similar to the actual prevailing cost of power when they were planned, there seemed little risk that they would not be an economic success given the background of a rising market price for power. Even if they did prove marginally more expensive than the alternatives, since they could constitute no more than 15 per cent of each REC's power purchases, and it seemed likely that any extra costs could be passed on to captive consumers, the risks appeared minimal. But it was the technology itself which did most to minimize the financial risk compared with conventional coal- or oil-fired plant. The construction time was typically 2–3 years, compared to 6–8 years for a coal-fired plant and the capital cost per installed unit of capacity was about half that of a coal-fired plant. This dramatically reduced the financial exposure of the companies especially prior to plant start-up.

CCGTs are largely factory-produced with much less site construction work than a coal-fired plant. This makes the control of capital costs much easier as the costs are mainly in the hands of the equipment suppliers rather than a number of civil engineering contractors. On this basis, the equipment suppliers were prepared to supply the equipment on a fixed-cost 'turn-key' basis with no scope for cost escalation while the project was being executed. Gas is a uniform fuel with few impurities and needs little of the expertise required for large-scale coal combustion. This had two consequences: the RECs did not need to be concerned about their inexperience in operating generating plant and equipment suppliers were more inclined to offer firm guarantees about the performance of the plant, backed up by insurance cover, because there appeared to be little scope for operator errors leading to breakdown.

Gas is a 'clean' fuel with major advantages over a coal-fired plant in terms of emissions of acid and greenhouse gases. The plant is also not very visually intrusive and pipeline delivery means that CCGTs need not be sited near rail facilities or a sea port. So the process of obtaining planning approval was not expected to be contentious.

These advantages made CCGTs an irresistible opportunity for the RECs and almost all signed up to build plant to fill a large proportion of their 15 per cent quota within two years of Vesting Day. But this so-called 'dash-for-gas' was halted almost as quickly as it started, partly because there was little scope left in their quotas for more plant, partly because the British Gas tariff under which the gas for most of the CCGTs was purchased was withdrawn and partly because of the political furore (described in Chapter 6) triggered by the rapid run-down of the British coal industry.

Despite the large volume of plant ordered, their contribution to competition has been small. Their 15-year power supply contracts effectively mean that the CCGTs do not compete in the Pool. They have to place a successful bid to be used, but the contractual cover means that the plant can be bid at zero cost just to ensure the plant is used. The income of the generating company is based solely on the terms of the power supply contract. The CCGTs represent a segment of the generation market that appears to be insulated from any competitive forces for 15 years and, since the buyer and the seller of power are one and the same, there is little incentive to minimize the cost of generation. These plants are effectively captive to the RECs to which their output is contracted and they compete neither in the mid-merit nor, except in a limited sense, in the base-load market.

THE NUCLEAR SECTOR

The main 'unfinished business' from privatization was the nuclear sector. After Vesting Day, the nuclear power plants were to remain in state ownership through Nuclear Electric, at least until completion of a Government review of its policy towards nuclear power which was scheduled for completion in 1994. The proposal that eventually emerged when the Review[3] was published in 1995 was that the Magnox plants would stay in state hands, in a new company, Magnox Electric. This company would be controlled by British Nuclear Fuels (BNFL), the state-owned nuclear fuel cycle and waste disposal company. BNFL already operates two Magnox stations which were built mainly to provide weapons-grade plutonium, but which are now primarily electricity generators. The two Scottish Advanced Gas-Cooled Reactor (AGR) stations, the five English AGR stations and the Sizewell B Pressurized Water Reactor (PWR) would be combined into one new company, to be called British Energy. Magnox Electric and British Energy were created on April 1, 1996, and the Government plans to privatize British Energy in 1996.

There was little prospect that the Magnox plants could be privatized because of their high generating costs, the limited operating life left and the huge costs associated with the closure and decommissioning of the plant. However, the vastly improved operating performance, the relatively low operating costs and their low average age meant that the AGRs and the new Sizewell B PWR could potentially survive in the market without subsidies on the operating costs. If a sufficiently low valuation was placed on the plant and other difficulties were ironed out, such as the provision of Government guarantees to cap financial risks including those resulting from accidents, escalation in waste disposal costs and changes in regulatory requirements, the privatization of at least some of the AGRs and the PWR was thought to be feasible.

It is important to stress (as argued in Chapter 7) that the economics of a new nuclear plant are so poor that there is little realistic prospect of any additional nuclear power plants being built whatever solution is chosen for them. In late 1995, Nuclear Electric withdrew its application for planning consent to build two further PWRs at Sizewell and announced it would not pursue the planning consent it already had to build a PWR at Hinkley Point.

The specific issues surrounding the nuclear plants are discussed in Chapter 7, but here we are concerned with the future contribution nuclear power plants might make to increasing competition in generation. It seems most likely that the costs associated with the Magnox plant, notably the closure and decommissioning costs but also the operating costs, will continue to be too high to allow them to compete in the Pool. They will therefore continue to operate as at present, as Pool price-takers, contributing little to competition. The high cost of running them will be borne by the consumer and the Government is committed to meet the closure costs. There will therefore be a strong incentive for the Government to allow them to continue to operate for as long as possible to postpone the need to incur expenditure closing and decommissioning them. Effectively this means that any competition between generating companies to build replacements for the 3000 MW capacity of the Magnox plants will be postponed for some time yet.

Although Scottish Nuclear and much of Nuclear Electric have been merged to form British Energy, the limited extent of the connections between the Scottish and English systems means that the Scottish plant will probably continue mainly to supply the Scottish system and the rest of the plant will supply England and Wales.

If the sale of British Energy proceeds, the level of debt imposed on the new company will probably be such that the cost of generation by the AGRs and the PWR, including servicing this debt, will be comparable to those of competing plant. This should prevent the new company making a loss but it should also ensure it does not make an embarrassingly large profit. However, given the lack of any realistic prospect of new nuclear plants being ordered, the important issue is the effect the new company will have on competition.

From the point of view of stimulating competition, an important issue is the extent to which British Energy competes in the Pool and the extent to which it competes for long-term contracts. It would be better in some respects if the new company was exposed to a significant extent in the Pool but it would be unreasonable to impose more onerous competition conditions on British Energy than on the existing generators. Given the inflexibility of nuclear plants, the nuclear plant will have to be given 'must-run' status. It therefore seems inevitable that the nuclear power plant will continue to be a Pool price-taker, not a price-setter and the direct contribution to competition in the Pool can only be small.

Privatization and the loss to British Energy of the income from the fossil fuel levy (FFL) will allow British Energy to compete more aggressively in the contract market and reduce its actual level of exposure to Pool prices through contracts for differences. However, to compete effectively for REC contracts, British Energy will need to diversify its stock of generating plants and bring in some mid-load and peak-load plants. Contracts for base-load power, as is provided by nuclear plants, command only a low price, and it is mid- and peak-loads that RECs are keen to contract for and which can command a premium contract price. British Energy could diversify in three ways: by purchasing old coal-fired plant from PowerGen, National Power or even Scottish Power; by building new generating plant, which would inevitably be of the CCGT type; or by purchasing existing CCGTs probably from the REC-owned independent power producers (IPPs).

The existing generators are unlikely to be enthusiastic about providing old coal-fired plant to increase the power of a new competitor and such a sale might only be possible under the direction of the regulator. The option of building new CCGTs would be more straightforward to carry out. Several of the nuclear stations to be privatized are sited in regions which are short of generating capacity and the connection charge to the national grid would be low or even negative – a new plant in the South West would tend to reduce the flow of power in the grid, reducing transmission losses and justifying a negative connection charge. However, the economics of building new plants to supply the mid- and peak-load are not likely to be very attractive. The time needed to obtain consents and finance, and to build the plant, would mean that there would be a significant delay before British Energy had sufficient plants in operation to start to compete as a supplier for the range of load factors.

Purchasing existing CCGT capacity from an REC may prove possible. The gas price has fallen significantly and the efficiency of CCGT technology has improved markedly since the 'dash for gas' with the result that some of the CCGTs are no longer so financially attractive as when they were built. However, the economics were so favourable for the first year or two, that some may already have paid for themselves. In these circumstances, some of the RECs may decide not to retain an interest in the generation business and be prepared to sell their CCGTs.

If the RECs' franchises are removed in 1998 and the supply market becomes fully competitive, the RECs' market shares will be at risk and this will shorten the length of contract the RECs will be able to sign. If this also increases the proportion of their power needs the RECs buy from the Pool, it may increase the pressure for the Pool to become a more realistic market. The shortening of contracts and the addition of a major new competitor should mean that the market for contracts will become significantly more competitive, especially if a nuclear company is competing from a reasonably comparable cost base.

The regulator will doubtless welcome these possible gains in the competition in the generation market, but in other respects the emergence of British Energy as a major generator may be less welcome to him. On its privatization, British Energy would become about as large as the largest existing generator, National Power, and the regulator is known to have argued strongly that the nuclear plant should be split between two privatized companies in order to create more competition.

Barring unforeseen circumstances, British Energy's nuclear power plant would all continue in service for at least a further ten years and any acquisitions of gas- or coal-fired plant would make the new company even larger. The decision to force National Power to sell some of its coal-fired plant suggests that the regulator is already uncomfortable with its size. He may therefore not look kindly on plans of British Energy to build gas-fired plant and he would certainly not be inclined to intervene to allow it to acquire existing coal-fired plant.

The option for British Energy to integrate into the supply business appears to have been closed for the time being by the Government's decision not to allow the takeovers of RECs by National Power and PowerGen.

THE POWER POOL

THE POOL IN THEORY

The Pool was the main arena in which competition in generation was expected to take place. As discussed in Chapter 3, the original objective for the Pool was to have a two-sided market with generators placing bids of the minimum price at which they were prepared to supply power, and customers placing bids of the maximum price at which they were prepared to purchase. The plans for this 'double pool' had to be abandoned about six months before Vesting Day because the computational requirements proved too severe. The short period remaining meant that the old CEGB dispatching software had to be adapted to operate as a one-sided Pool with only generators placing bids. The Pool mechanism is operated by a non-commercial organization known as the Pool which is staffed by NGC personnel.

The rationale for allowing only plants that have placed successful bids into the Pool to operate was that this appeared to provide non-discriminatory access to the grid. It was feared that the alternative, that contracted power did not need to bid into the Pool, would have effectively given priority on the grid to contracted power and, where grid capacity was tight, could have been used to force 'spot' power, no matter how cheap, off the system.[4]

The essential logic of the Pool is simple, although the detailed workings are complex and probably understood by only a handful of people. All generators that wish to operate their plant must place a bid with the Pool that applies to each half hour of the next day. The Pool software matches the bids against forecasts of demand choosing the lowest bids necessary to meet demand. The Pool Purchase Price (PPP) for each half hour is largely set by the price bid by the highest successful bidder, the System Marginal Price (SMP).

The major addition to the SMP is a term which relates to the amount of capacity that is bid that is in excess of that required. The closer supply and demand are, the higher the payment due under this term will be. Formally, the addition is calculated by the formula VOLL multiplied by LOLP, where VOLL is the value of lost load and LOLP is the loss of load probability. VOLL is intended to be a measure of the economic cost of not supplying 1 kWh of electricity and was set at £2/kWh.[5] While there are clearly some costs associated with not supplying power, it is not at all clear how these costs should be estimated. How the figure used was derived has not been established, but £2 is suspiciously close to the figure derived from dividing gross domestic product by total electricity consumption. Thus, it is approximately the average amount of national economic output produced per unit of electricity the nation uses.

LOLP is the probability that the amount of power bid will prove insufficient, for example due to capacity shortage combined with plant breakdowns or demand forecasting errors, to meet demand. If there is insufficient capacity to meet demand and disconnections are unavoidable, the Pool Purchase Price paid to all successful bidders will rise to the value of lost load, in other words, it will be about £2. LOLP is only significantly greater than zero when supply and demand are very close and it is only then that this financial addition to the highest successful bid price will be significant. The rationale for this term was that it would provide an economic incentive for the owners of peak-load plant to maintain their plant in service and to build additional capacity when required. Peak-load plant to is only used for a few hours each year and it was assumed that, without this addition, the Pool price for these few hours would not provide sufficient income to meet the fixed costs of maintaining peaking plant in service. If the system is properly balanced, the VOLL/LOLP mechanism should therefore provide sufficient income for the marginal peak load plant to repay its fixed costs.

The price at which power is purchased from the Pool, the Pool Output Price (POP), is the Pool Purchase Price plus an uplift term which covers the cost of grid services such as transmission losses, reactive power and the cost of stand-by capacity or spinning reserve.

There are a number of other features to the system to take account of inflexible plant, that is plant that cannot be turned on or off at half-hourly intervals because of its physical characteristics. There are also mechanisms to deal with grid bottlenecks. If a bid is placed for a plant that is below system marginal price, but the plant cannot be used because of shortage of grid capacity, the plant is paid the imputed profit it would have been made had it operated. This is calculated as the difference between its bid price and the Pool Purchase Price.

The Pool in Practice

The main problems that have emerged with the operation of this compulsory Pool have resulted from the dominance of National Power and PowerGen in Pool price-setting. This dominance exists because they own all the plant (old coal- and oil-fired stations) which is only required intermittently (on mid-merit load). National Power and PowerGen are able to predict with some precision which Pool bids will be at the margin and will therefore set the Pool price. This is paid by the Pool to all plant for which a successful bid has been placed regardless of how low the actual bid was.

Contracts for Differences

The fact that more than 90 per cent of power bought and sold has not had to place a realistic bid into the Pool, because the price paid by the purchaser to the generator ultimately has been determined by a contract, has disconnected the bidding system from costs. Any plant covered by contracts which a generator wanted to use has been able to place a zero-price bid, yet has still received the contract price regardless of the Pool price. All the operating IPP plants and most of National Power and PowerGen's plants have been covered by such contracts, while Nuclear Electric's plants have been purely passive price-takers which effectively cannot set the Pool price.

This has left National Power and PowerGen with power to set the Pool price at all times except for the limited periods when bids for the pumped storage plants were at the margin.[6] This power over the Pool price has given them scope to force Pool prices up or down in the knowledge that most of their income was covered by contracts for differences which were independent from Pool prices. They have been able to make excessive profits with the marginal plants which set the Pool price and which were not covered by contracts by bidding them at well above economic cost. This dominance has made it highly risky for a new generation company to build a plant which relies mainly on the Pool for its income because its plant could be forced off the system by predatory pricing by the two large generators. It also means that the option for RECs and final consumers to buy power from the Pool directly has been seen as risky to them, partly because of the volatility that Pool prices have shown and partly because Pool price could be manipulated by two of the players on the supply side of the system. Few large final consumers of power have purchased power from the Pool and one of these, the chemicals company ICI, has become so dissatisfied with its experience that it plans to build its own generating plant. The RECs are also showing little sign of increasing their direct purchases from the Pool.

This duopoly power, and the protection that contracts for differences gives them, also discourages the emergence of a third-party, wholesale electricity market with traders, which do not own generating plant, buying and selling power, for example in a futures market, in the way that other commodities are traded. In the current situation, National Power and PowerGen will always have a strategic advantage over a trader through their ability to set prices, and a trader could always be forced out of the market by price manipulation.

Regulatory Measures to Reduce Duopoly Power over the Pool

The regulator was increasingly concerned about what he saw as excessive profit from the Pool by National Power and PowerGen and the fact that the duopoly power over the Pool they started with has not been reduced at all by new entrants. As discussed above, he negotiated an undertaking in 1994 from National Power and PowerGen that they would use their best endeavours to sell 6000 MW of coal-fired plant. A second part of this deal was that they would ensure that the Pool price would be held at an annual average of about 2.55p/kWh for two years from April 1994.

This price cap did not make the Pool a predictable market that buyers and sellers can put their trust in. National Power and PowerGen only committed themselves to maintain the Pool price at annual averages of 2.55p/kWh weighted by time and by demand. This has allowed the volatility of Pool prices actually to increase after the imposition of the cap. This has been demonstrated by the fact that the standard deviation of Pool prices in 1994/95 (which measures how variable Pool prices are) was nearly four times larger than it was in 1993/94. A volatile Pool price has significant advantages for the generating companies. It discourages customers from buying directly from the Pool, because of the uncertainty of prices, and favours the option of buying directly from generators at fixed prices under contracts for differences. These contracts for differences give generators a more predictable, risk-free stream of income than competing in the Pool would do.

The impact of the price cap on the regulator's attempt to break the power of the duopoly has been counterproductive: the objective of encouraging new generators to enter the market by building plant that would compete genuinely in the market has been severely set back. The price cap was the strongest possible signal to potential investors in generating plant that the Pool was not a genuine market and that it was vulnerable to arbitrary and unpredictable interventions by the regulator – not the sort of environment to entice entrepreneurs.

The capped Pool price was a reasonable reflection of the costs incurred by the duopoly with the marginal plants that are actually setting the Pool price, that is, their older coal plants which are fully depreciated. But, it was well below the level necessary to provide a commercial return on capital for new investment. Excess profits for National Power and PowerGen may therefore have been curtailed but the incentives for new entrants have been further reduced. In addition, much of the capacity built by RECs burning gas now appears uneconomic compared to the capped Pool price – the consensus is that they produce power at a cost in early 1996 of 2.8–3.0p/kWh. Their contract protection and the inertia of their customer bases mean that the RECs will probably not lose money on this investment. However, they will not be encouraged to repeat the experience of building generating plant in competition with National Power and PowerGen, especially if the protection of long-term contracts, which has limited their losses on this occasion, is not available. The regulator decided in January 1996 that the Pool price cap would not be renewed when it expired in April 1996.

Security of Supply

The winters of 1994/95 and 1995/96 also saw some extreme peaks in the Pool price for reasons that often seemed to have little to do with the cost of supply and demand and which raised question marks about the security of supply in the new system. Despite the duopoly's assurances on Pool price, the average Pool selling price for December 1994 was 4.2p/kWh and the price peaked at over 60p/kWh. The reasons behind this were not weather-related: peak demand (46 GW) was 2 per cent below that of the previous year. The main factor was the breakdown of two of the nuclear power plants, reducing available capacity by about 2.5 GW. This diminished the margin of spare plant down to minimal levels and, had demand been as high as the previous year, the lights might well have dimmed or gone out. To compensate for these high December prices, National Power and PowerGen ensured that Pool prices were extremely low in February 1995 so that they would meet their commitment to the regulator to maintain the annual average Pool price at the agreed level. As a result, the annual averages were close enough to their targets to satisfy the regulator.

Again in December 1995, Pool-price levels rose to a new record of more than £1/kWh, due to a coincidence of cold weather, strikes in France leading to a loss of the imports from France, which are generally about 2 GW, and break-downs at a number of nuclear plants. In January 1996, there was serious concern that there would not be sufficient capacity to meet demand. This time the problem was availability of gas for power generation. Most of the IPP CCGTs are supplied with gas under interruptible contracts signed with British Gas. British Gas needs to 'interrupt' power-station gas supplies on cold winter days in order to ensure that the high demands that domestic consumers make then can be met.

Nuclear and coal-fired plants do break down, strikes do happen and weather a good deal colder than in the last two years does happen. The failure to anticipate the inevitable consequences of IPPs being supplied under interruptible gas contracts is particularly difficult to understand. If the loss of small amounts of plant puts the entire system in jeopardy, questions need to be asked about the integrity of the system under the new competitive regime.

Because the generation market is expected to become a fully competitive market in which several companies compete, it is difficult to impose mechanisms which oblige one of the competitors to ensure there is a sufficient reserve margin. The basic problem is that the criterion for a generator in deciding whether to retain or retire a power station (or whether it decides to build new capacity) is not whether the capacity will be needed to keep the lights on, but whether it will make a profit. If a plant is not expected to operate profitably, the generating company has an obligation to its shareholders to retire it (or not build it). If as a result the lights go out, no blame can be attached to any one of the generating companies.

VOLL/LOLP Payments

The VOLL/LOLP mechanism to provide financial incentives for generating companies to build and maintain peak-load plant has proved naive. The level of the peak load is highly dependent on how cold the coldest day of winter is. For example, over the past decade, demand has, on a number of occasions, been 2–5 per cent higher or lower than is estimated would have been the case in an average winter. Even if the amount of peaking capacity is appropriate and the scale of VOLL/LOLP payments is accurately calibrated against the costs of maintaining peak-load plants in service, a run of warm winters would mean that peak load plant would be little used and owners of peaking plant would incur heavy losses causing them to retire the plant. At worst, a shortage of peaking plant could mean that on cold days when demand was highest, there would be insufficient capacity to meet demand. If there was a compensating surplus of mid- or base-load plant, a shortage of peaking plant could raise overall costs because of the need to use mid- or base-load plant, which is generally not designed to operate in short bursts, as peaking plant.

Another problem with the VOLL/LOLP payments is that with only two companies generally setting the price at the margin the mechanism is vulnerable to abuse. There is little reason, in seasons such as summer when demand is low, for a generating company to incur the cost of ensuring that a plant, with no realistic chance of being required for months at a time, is ready for service and a bid for it placed with the Pool. For entirely legitimate reasons, the gap between demand and available supply at any one time will therefore often tend to be small. In this situation, the risk is that by choosing not to bid some additional plant, National Power and PowerGen could engineer a supply/demand balance which would generate a large VOLL/LOLP payment which would be paid to all successful bidders. Much of their plant would not profit from this because of the contracts for differences, but useful extra income would be generated on the capacity directly exposed to Pool prices and the message that Pool prices were unpredictable would be reinforced. The regulator strongly suspected National Power and PowerGen of such exploitation in 1991 and warned them not to persist with such behaviour. As a result of the revealed flaws in the VOLL/LOLP payment scheme, the regulator has signalled that he expects the VOLL/LOLP mechanism to be replaced by a more effective method of dealing with the problem of giving incentives to maintain peak-load plant.

Practical Pool Problems

After the December 1994 price spikes, there were further leaps in price in March and April 1995, to 20p/kWh in March and 80p/kWh in April. Unlike the December 1994 and December 1995 price spikes, much of which were accounted for by VOLL/LOLP payments, the spike in April had nothing to do with any shortage of capacity. It appears to have been mainly the result of a weakness in the Pool software.

SUMMARY OF POOL EXPERIENCE

The credibility of the Pool is now at a low ebb. A projected re-writing of the Pool computer software and a new system to replace the VOLL/LOLP payments might solve some of the practical problems. A break-up of the duopoly power of National Power and PowerGen might reduce the scope for monopoly abuses. However, such steps would not remove the doubts about the ability of a universal Power Pool in the form chosen to ensure that the national stock of generating plant is of appropriate size and composition. If this is not the case, costs to final consumers will not be minimized and security of supply cannot be guaranteed.

COMPETITION FOR FINAL CONSUMERS

SUPPLY COMPETITION IN THEORY

Giving consumers the power to choose their electricity supplier was a high priority for the Government when the industry was privatized. Previous utility privatizations were becoming unpopular as consumers (and some politicians) saw them as just a transfer of a monopoly from the public to the private sector. The benefit to consumers of being able to shop around for the cheapest deal and to withdraw their custom if they were dissatisfied with the service they received was therefore an important element in selling the electricity privatization plans to the public.

If the new system was operating as the Government envisaged, competition in supply should not be an important direct mechanism for promoting economic efficiency. Of the other elements in the consumer price, transmission and distribution will remain regulated monopolies and will be charged for at standard non-discriminatory tariffs. Generation should be a competitive market driven by the Power Pool, through which all generators must sell their power. If the Pool sets the price of wholesale power, either directly by consumers buying from it or indirectly by setting the price at which contracts are agreed, it will be difficult for one electricity supplier to purchase its power more cheaply than any other. This leaves only the direct costs of supply, such as meter-reading and billing, which make up no more than about 5 per cent of a typical electricity bill, on which suppliers can establish an economic advantage over their competitors (Table 4.2).

Table 4.2 Cost breakdown for three annual bills

	Generation	*Distribution*	*Transmission*	*Supply*	*FFL*
<0.1 MW (£405 pa)	52	26	5	7	10
0.1–1 MW (£42,000 pa)	59	22	5	4	10
>1 MW (£475,000)	69	15	5	1	10

Source: Centre for the Study of Regulated Industries (1995) *The UK Electricity Industry: Charges for electricity services 1995/96*, Chartered Institute of Public Finance and Accounting, London, p 13.

However, there are also indirect ways in which competition in supply might have efficiency benefits. Competition should ensure that the price paid for any service will tend to reflect accurately the costs incurred in supplying it. If the price is too high, there will be scope for new suppliers to enter the market, driving the price down and the effect of these price cuts will have to be recovered from consumers for which the price does not fully reflect the costs. Advocates of a competitive system frequently complained that the tariffs operated under the old system contained such cross-subsidies. Cross-subsidies are anathema to free-market enthusiasts who believe that they promote wasteful and inappropriate patterns of usage at the expense of the performance of the economy. For example, if energy-intensive industrial consumers pay too much, their international competitiveness is adversely affected, and if consumers in general pay too little, they tend to under-invest in energy efficiency and hence waste electricity.

The main beneficiaries of cross-subsidy in the old system were seen to be the low-volume domestic consumers because the fixed or standing charge did not fully reflect the non-variable cost of supply. This group of consumers includes the poorest in the community. Increasing fixed charges to them while the industry was in state-ownership would have been politically damaging: it would have been portrayed as preventing their access to a basic necessity. The balancing reduction in unit charges would also have tended to increase consumption, in conflict with the aim of promoting energy efficiency. The largest consumers also received preferential treatment under long-standing deals, sometimes explicitly subsidized by Government, which were meant to improve their competitiveness in international markets.

As for all industries based on a common network, allocating costs between different groups of consumers for distribution and transmission is an art based on judgement, rather than a science and, for example, there are wide variations internationally on how transmission services are charged. However, these are not competitive areas of the business and there is no obvious reason why the owners of the network should discriminate for or against any one set of consumers. There can never be a 'right' way of allocating joint overhead costs between different markets: it will remain a matter of judgement and probably of controversy in which the regulator will have a keen interest.

For supply, in which competition was meant to operate, cost allocation problems are much less acute although by no means negligible. However, while the Government's reforms were meant to make trade in electricity much more like any other consumer purchase than in the past, it would have been politically risky to remove all the layers of protection. It would not have been acceptable if, for example, poor pensioners who consume very little power and sometimes get into difficulties in meeting their bills either paid a very high tariff or were unable to find a supplier willing to supply them. Particularly when competition is introduced to the domestic sector, the dangers of 'cherry-picking', whereby a new supplier targets only profitable consumers leaving the incumbent supplier to shoulder the extra costs of their remaining consumers, must be guarded against. To do so effectively is difficult, as the gas regulator has begun to find out even in advance of competition in the domestic gas market.

Supply Competition in Practice

In practice, giving choice to all consumers could not immediately be accomplished and the difficult issues raised above have yet to be faced. As discussed in Chapter 3, an immediate opening of the entire market would have had an adverse effect on the level of proceeds from the sale, because of the extra commercial risk the incumbent suppliers, the RECs, would face. In addition, the existing distribution companies were inexperienced at operating in competitive markets and new entrants were likely to emerge only gradually. The decision was therefore to open the market up slowly. Only consumers with a maximum demand of greater than 1 MW were allowed choice of supplier from Vesting Day in 1990. The limit was reduced to 100 kW in April 1994 and it is scheduled to be removed completely in April 1998. Customers in the competitive part of the market could choose from any of the licensed electricity suppliers. In practice, this initially meant the 12 RECs and the two privatized generation companies, but it was hoped that other companies would enter the field as competition developed.

It was always likely that opening the market for 1 MW consumers would generate considerable competition. This was due to the small number of large consumers and their high annual bills (typically £0.5m) which meant that marketing efforts could be carefully targeted and were cheap.

Although revenues from such customers are high, the fixed costs of supplying them are low, and supply costs typically make up only about 1 per cent of a bill. Such customers are large enough organizations to negotiate effectively and will have no supplier loyalty – if their power supplier is not offering the best deal available, the customer will soon switch.

Nevertheless, in the period immediately after Vesting Day, it was important that the newly privatized companies established themselves strongly, and successfully competing for large customers could be used to good public-relations effect. As shown above, the pricing formulae for the RECs set at Vesting Day were not very demanding and any losses in this sector of the market could be offset by profits from the franchise markets and from the operation of the distribution network.

Since then, the scope for non-price competition has become clearer. Suppliers can offer services such as giving information about electricity demand patterns which allow the company to save energy or to move consumption away from periods when power is expensive. Large consumers are usually finding that their suppliers are now much more flexible and more willing to provide services tailored to their individual needs than in the past.

By 1994, when the franchise limit was reduced, both sides of the market were more sophisticated and competition was no less strong than it was for the 1 MW market. The first year was marred by difficulties with the new data collection, metering and billing systems necessary to allow competition; even so, about half of the consumers in this sector had changed supplier after one year (see Table 4.1).[7]

While it is clear that competition in the non-franchise market has been vigorous and generally to the benefit of customers in this sector, few new suppliers of power have emerged. Some commentators had expected that sophisticated trading markets, for example in futures, would have developed as for other commodities. However, traders may well have been deterred by the dominance that National Power and PowerGen can exert over the Pool price.

Another factor behind the lack of emergence of new suppliers may be the low profitability of the large-consumer sector of the supply business. In the non-franchise market, it is thought that the suppliers do little more than break even. Whether the generators and RECs choose to cross-subsidize their supply businesses from generation and distribution respectively or whether they are locked into a competitive spiral from which they cannot escape is not clear, but low profitability is a powerful disincentive to new entrants.

PROSPECTS FOR COMPETITION IN THE DOMESTIC MARKET[8]

The introduction of competition for large and medium-sized consumers has not been without difficulties, but the difficulties are likely to be greater by far in the domestic (or residential) market. The main problems appear to be the high cost of the sophisticated meters which were thought necessary for customers to switch suppliers, the high marketing cost of winning new customers and the low level of profits generated by each individual consumer. On average, RECs make only £5 per customer per year.

Competition amongst larger consumers has required the use of sophisticated meters which transmit consumption data on a half-hourly basis to the supplier. This allows the supplier to derive tariffs which accurately reflect the hourly cost of supplying the final consumer and also generate useful data which allow the consumer to modify consumption patterns to reduce energy costs. The cost of such meters for domestic consumers is reportedly about £50 and there would also be huge data processing costs associated with dealing with the mass of data each consumer would generate. These costs are well above those most consumers could justify from the price reductions they could achieve or a supplier could justify incurring to win a new customer.

A possible way round this problem borrows from a scheme adopted in Norway, whereby the consumption pattern of a household is imputed from its characteristics, such as size and number of inhabitants. 'Consumer profiling' would remove the need for new meters but suppliers would not have any means of knowing what costs a given consumer incurs nor would the consumer have any incentive to modify consumption patterns to reduce costs. A major element of competition – accurate cost-based price signals to individual consumers – would be foregone.

The cost of winning new customers is likely to be high especially for a new company with no local reputation competing against a well-entrenched supplier. The large number of consumers makes advertising difficult to target and substantial guaranteed cost-savings would have to be offered to persuade consumers to change. From experience with other utilities, the operational cost of acquiring a customer is £25–50 and an annual discount, also of £25–50, would also be necessary.[9] These costs could be lower if profitable customers are targeted, 'cherry-picking', or if other utility suppliers, such as water and gas companies chose to move into the electricity business. The most obvious partner is a gas company but a rough estimate of the savings from supplying electricity and gas to the same customer is only £6–£8 a year.[10]

Given that the average annual domestic consumer bill is about £400, of which supply accounts for about 7 per cent, that is about £30, it is difficult to see how supply profits alone could justify the marketing and discount costs of winning new consumers. This does not augur well for the prospects of sustainable competition by new entrants without special advantages, such as monopoly profits from other businesses such as water or gas supply.

CONCLUSIONS

The conditions imposed by the Government over the first 3–5 years of operation of the new competitive regime meant that the direct impact of competition could only be small (see Table 4.3 for a breakdown of the movements in customer bills since 1990). Contracts between generators and RECs and the retention of most of the RECs' franchise privileges gave little scope for competitive forces to have an impact. Nevertheless, this period has given time for markets to take shape and for competitive mechanisms to emerge.

In the generation sector, the main problems have been that the structure of the generation market is too concentrated, the Power Pool has not proved effective, and the concept of a universal daily Power Pool may not be compatible with the standards of security of supply demanded by consumers.

The problems with the structure of the generation sector appear to be attributable to the efforts to privatize the nuclear sector which left only two large dominant generation companies. New generation companies emerged faster than most people expected, but they provided little additional competition. This is because of the contractual protection the CCGTs these companies own were built under and because these CCGTs do not compete in the mid-load and peak-load parts of the market. The regulator has sought to improve the structure by forcing the two large privatized generation companies to sell some capacity and by requiring the NGC to sell its generation business. However, these gains may be counterbalanced by market-driven re-organization, in particular the continuing pressure to reintegrate generation and supply.

Table 4.3 Typical Electricity Bills, 1990/91–1995/96 (£)

	1990/91	*1993/94*	*1994/95*	*1995/96*
Domestic, standard tariff	262	291	286	279
Domestic, low use	105	117	116	111
Domestic, day/night tariff	350	390	387	378
Non-domestic, small	1,506	1,666	1,635	1,608
Business, medium	12,408	13,890	13,764	13,542
Business, large	150,355	165,204	165,232	163,821

Source: Centre for the Study of Regulated Industries (1995) *The UK Electricity Industry: Charges for Electricity Services 1995/96*, Chartered Institute of Public Finance and Accounting, London, p 13.

Notes

1 Bills are expressed in money of the day. Inflation, as measured by the retail price index, between October 1990 and October 1995 was 16.1%

2 Domestic standard tariff consumers are assumed to consume 3300 kWh per year; low-usage domestic consumers 1000 kWh per year, day/night domestic consumers 3000 day kWh and 3600 night kWh per year, small non-domestic consumers, 20,000 kWh per year, medium business consumers 175,200 kWh per year and large business consumers 2,628,000 kWh per year

It is far from clear whether the proposed privatization of most of the nuclear sector will increase or reduce the competitiveness of the sector. The emergence of a third large generator may break up the duopoly power, but the sheer potential size of British Energy gives cause for concern.

The concept of a two-sided, universal Power Pool was an ambitious one, but one which was already compromised by the withdrawal of demand-side bidding before the new system was put in place. How far the problems with the Pool are due to the structural deficiencies in the generation side or to fundamental problems in the design of the Pool will become clear only if the generation structure is made more competitive. It may be that a generation market that is fully competitive on a day-to-day basis would not be a sufficiently secure market for potential investors in generating capacity to risk their money on. If this is the case, new capacity would be difficult to finance and the security of supply of the system would be unacceptable.

Despite teething problems in the 100 kW market, competition is now a reality for large consumers. The low proportion of consumer bills constituted by supply may mean that supply competition is of more symbolic than economic importance. If customers still face a monopoly supplier, they may feel that the changes to the system have brought little benefit to them. If competition does not emerge at all for domestic consumers, or if it benefits only large, profitable, domestic consumers, there is a serious risk that competition will be paid for by those least able to afford it: small, low-income domestic consumers living in areas where utility debts are high. Together with security of supply risks, these are the same concerns which have been widely expressed about competition in the UK domestic gas market. The fact that it is proposed that both markets will be fully opened to competition on the same day, 1 April 1998, compounds the anxiety.

5

Regulation

Gordon MacKerron and Isabel Boira-Segarra

INTRODUCTION

Regulatory issues, both economic and environmental, have a history as long as that of the whole electricity supply industry. Both have tended to take new and distinctive forms since the late 1980s. In the case of economic regulation, this has been a direct consequence of the privatization of the industry, whereas changes in environmental regulation have had different roots.

While economic regulation is the more obvious subject for this book, we also give a less detailed treatment to environmental regulation. There are three reasons for looking at environmental regulation. First, environmental regulation has become increasingly important, especially in its potential implications for the generating side of the industry. Second, environmental regulation gives powerful incentives to the regulated – incentives which may either reinforce or contradict the incentives given by economic regulation. Given that the two regulatory systems operate in almost total isolation from each other, such consideration is important. Third, it is often difficult to separate economic from environmental controls. For instance, the imposition of VAT on domestic uses of electricity was in an immediate sense an economic issue. However the Government justified the decision on the grounds that it would contribute to meeting national targets for the reduction of carbon dioxide emissions.

This overlapping of the economic with the environmental also illustrates a further important point about 'regulation'. Although the activities of the industry-specific regulator in the economic sphere and the designated agencies in the environmental area are vital, they are only part of a much wider regulatory framework. Regulation, as we use the term throughout this book, refers to all state-authorized control systems and decisions that affect the ESI, including those of Government departments and general economic regulatory bodies, principally the Office of Fair Trading (OFT) and the Monopolies and Mergers Commission (MMC). The issues of regulation in this broad sense are therefore in principle identical to all the pre-privatization issues of state control (outlined in Chapter 2), though they are now – in economic regulation – often exercised by different agencies, following different rule-books.

ECONOMIC REGULATION

THE INSTITUTIONS

The most obvious starting place for a consideration of economic regulation is the Office of Electricity Regulation (OFFER). It has a staff level of just over 200 (of whom about 90 work in its regional offices) and its annual budget of around £11 m is paid for by its licensees.[1] Strictly we need to talk of the Director General of Electricity Supply (DGES) because the Electricity Act of 1989 recognizes the DGES alone as responsible for the conduct of regulatory activity.[2] This one-person system of regulation has led to considerable criticism, and in the case of the gas industry to a significant change of regulatory style and to some extent substance when the regulator changed. This over-personalization of economic regulation in Britain has been the norm for all the privatized utilities. However, it does not represent universal British practice: for example, the MMC has always operated collectively, and for any one inquiry, five or six commissioners will be appointed.[3]

OFFER is an agency of the state but is not directly responsible to any Government department: it is a non-departmental agency. This reflects the intention that it should operate at arm's length from Government on a day-to-day basis. This does not mean that Government will not seek to influence the regulator (indeed as shown below, Government formally shares all the main regulatory powers with OFFER) but it does mean that OFFER does not take political 'orders' from Government, and has sometimes acted in ways which conflict with the immediate desires of the Government of the day. If OFFER is not really responsible to Government, the question remains as to whom or what it is accountable, and there is no clear answer here (see Chapter 11 for a more detailed discussion of the accountability issue). Parliamentary scrutiny is limited. Although Select Committees of the House of Commons may require regulators to appear before them, there is no regular, say annual, review of their activities, and consumers have no independent voice. The Regional Consumers' Committees established by the Electricity Act are actually appointed by OFFER, and consumer complaints are OFFER's direct responsibility.[4]

OFFER's responsibilities are wide-ranging, but it is important to note that the major regulatory duties are shared with the Secretary of State (ie the Government). The Electricity Act certainly gives the Director General sole powers to implement the details of regulation (eg amendment of the licences which all electricity companies must obtain) but on all major issues the responsibility is held jointly with Government rather than being vested solely in OFFER.[5] The widespread public perception that the major responsibilities of regulation are entirely the preserve of OFFER is therefore more a reflection of the present Government's style of 'rolling back the state' than a precise reflection of the underlying structure.

The primary regulatory responsibilities in the 1989 Electricity Act are to ensure that reasonable demands for electricity are satisfied; to license suppliers of electricity and ensure that they are financially viable; and to promote competition wherever possible.[6] Consumer protection does not figure directly in this list of primary duties (though it does constitute a secondary responsibility). This is not so much because consumer protection was thought unimportant by Government – indeed it is the dominant rationale for regulation – but rather because it was assumed that a competitive market in generation and supply was the best guarantee of consumer protection. This is a testable proposition rather than an axiom and a major theme of this book is to provide such testing of the 'automatic' virtues of competition in the industry as is yet possible.

A second important feature of OFFER's responsibilities is the primacy given to the duty to promote competition (in contrast to telecoms and gas where the duty is secondary). This characteristic of OFFER's role marks a real departure in the scope of regulation in electricity supply compared to that in other industries (or indeed in the same industry in other countries, including the USA, where there is a long history of utility regulation). Almost all of the economic theory of regulation is about how to control monopoly powers, and virtually none is devoted to how to promote competition.[7] The good reason for this imbalance is that it is inherently difficult to theorize in detail about the best methods for promoting competitive markets, and traditionally this task of ensuring competition has been seen as a role to be performed by a mixture of the political and legal system, rather than devolved to a one-person economic regulator.

Much of the criticism that OFFER has attracted – especially from within the ESI itself – has been of its attempts to determine the course of action that will best promote a more competitive market. This is not very surprising, given that the decisions the DGES has made have been potentially so profound for the future of the industry. An example is the decision to require the two big generators to sell off 6000 MW of plant to competitors. It is doubtful whether it is appropriate to leave decisions about the structure of an important industry to a single state appointee.

Further, other decisions about the future competitive structure of the industry have remained firmly in the hands of Government – for example the Cabinet decision to try to sell the privatizable parts of the nuclear industry as a single company, thus side-stepping the chance to open generation to further competition. This is a decision which contradicts OFFER's stated preference for a more competitive structure in the privatized nuclear power sector.[8] The uneasy sharing of the same role between the DGES and the Government at large suggests a need for reform (see Chapter 11). It also shows that the Government retains important and direct powers of economic regulation of the ESI outside the scope of the Electricity Act as well as within it.

The other main regulatory agencies that may be important for the ESI are those which are responsible in an economy-wide sense for ensuring that fair competitive practices are adhered to, and that monopoly powers are policed. These are the OFT and the MMC. The OFT is a first 'port-of-call' for those who feel that any trading practice has been unfair, and it has expressed views about the implications of proposed mergers involving the RECs. More obviously significant to the electricity supply industry is the MMC. It has no power to act independently – it can only act on a 'reference' from Government or from a regulator in defined circumstances – but it has extensive powers of legal and economic investigation and its recommendations carry political weight (though they are not, unless there is a clear breach of law, automatically enforceable).

The MMC can in principle play at least three roles in utility regulation.[9] The first, specific role is that of adjudicator of disputes between the regulator and the regulated. When OFFER announces proposals for new price controls (and thus licence amendments) it is open to any of the affected companies not to accept OFFER's proposals. In such an event, the regulator must call in the MMC to adjudicate. However, the MMC can in principle not only accept OFFER's original formulation, it can make it even more stringent on the companies, and further, the DGES has a major role in framing the MMC terms of reference. Objections by companies are not made lightly, and to date only one such objection has been made – by Scottish Hydro-Electric against the 1994 OFFER proposals for changes to distribution and supply charges.

The second potential role for the MMC lies in its general powers to investigate monopoly. British competition law allows the MMC to investigate potential abuses of monopoly power if any one company holds 25 per cent or more of the share in a defined market. Since privatization, the two incumbent generators have consistently held market shares above 25 per cent and this allows either the Government or OFFER to make a reference to the MMC for such an investigation at any time. This is a substantial reserve power for OFFER to hold in the generation market, because the MMC could recommend the breaking up of the generators into a larger number of companies in the course of such an investigation – a potential outcome the affected companies may be expected to avoid if at all possible.

Third, the MMC can also in principle be involved in adjudicating the merits of proposed takeovers and mergers in the industry, such as began to be proposed in the area of distribution and supply in late 1994 and 1995. In such cases the reference must come from Government, and while both OFT and OFFER have expressed misgivings about all the mergers, the Government has chosen to send only two proposed mergers (those involving National Power and PowerGen, and leading to vertical re-integration) to the MMC (see Chapter 10). Thus while the MMC has only been involved directly in the regulation of the ESI to a limited extent so far, it is a powerful ghost.

Finally the European Commission, especially through its competition Directorate, is also a potentially important influence on the economic behaviour of the industry. However, because the England and Wales reforms are in the vanguard of the direction in which the Commission is trying to push other member states – broadly towards more liberal and open markets – there have in practice been very few issues where the Commission has had any direct impact. The only occasion where the Commission has yet played an important role in ESI regulation has been in nuclear power, where the British Government was told that the fossil fuel levy and NFFO (see Chapters 8 and 9) could last only until April 1998.

GENERAL ISSUES OF REGULATION

The regulation of large and strategic industries like electricity, fundamentally affecting the well-being of all citizens (and voters) is never likely to be a wholly economic affair. As Chapter 11 argues, regulation needs to achieve a balance between the interests of the main stakeholders in the industry (especially consumers and shareholders), and cannot ignore social issues entirely. In the privatized British system, regulation has been rather narrowly focused: apart from protection against monopoly abuses, it has concerned itself almost entirely with issues of economic efficiency, which are themselves held always to be advanced by the promotion of competition. Regulatory practice by OFFER has had limited interest in wider issues within its own explicit remit, including energy efficiency, health and safety, research and development and the environment.

The England and Wales case is a particularly interesting test of the new forms of regulation that accompany privatization and liberalization. As the most complete example of liberalization-with-privatization, England and Wales offers an opportunity to see how far radical market-based reforms can lead to a genuine reduction in the role of the state in the industry – in simple terms, 'can privatization lead to real deregulation?'. Clearly the Government expected the initial regulation of the industry to be light, and believed that it should get even lighter over time. The thinking here was simply that there would be a direct trade-off between regulation and competition, and that as competition advanced, so regulation would retreat. This retreat has clearly, as we show below, not taken place, even though competitive forces are undoubtedly stronger in early 1996 than they were six years earlier. We offer some explanations for this outcome in the conclusions of this chapter.

As OFFER has been virtually the only agency responsible for direct regulation of the ESI, the discussion here inevitably focuses mainly on its activities. The regulation of monopoly powers has been the staple activity of OFFER and it has also attracted much criticism, mostly for its supposed leniency towards the industry. Professor Littlechild (the DGES) was one of the inventors of the system of price control for monopoly[10] that he is now, in the case of electricity, responsible for implementing. The principal weapon of control is a form of price regulation generically known as RPI–X, where RPI is the retail price index and X is an efficiency improvement term set and re-set by the regulator. The idea is that regulated prices charged by the industry should normally rise less rapidly than prices in general by an amount that reflects efficiency improvements deemed achievable in the technical and economic conditions of the industry (and reflected in the size of the X term). This general formula applies to all the main privatized utilities – gas, water and telecoms as well as electricity. An important issue has been the period over which price controls will apply. In a desire to allow the companies a relatively stable framework for planning, the price controls have generally been fixed for four- or five-year periods – a duration which puts pressure on regulators to get the X value 'right'.

The price control system was chosen in preference to other models of control, particularly rate of return regulation as widely practised in the USA. The disadvantage of rate of return regulation is supposed to be that there is an incentive for companies to inflate their asset base unduly (a given rate of return yields a higher absolute profit if it can be earned on a higher value of assets) and that it gives no direct incentive for companies to reduce costs as there is no mechanism by which companies can keep part of any efficiency gains for themselves.[11] By contrast, price control regulation is said to have a number of advantages. First, it is expected to be transparent and easily comprehensible to consumers. Second, it supposedly minimizes information requirements, as it appears not to require knowledge of asset values, rates of return etc. Third, it allows companies to keep the benefit of any efficiency gains greater than the X term, and thus gives incentives for efficient behaviour.[12]

However, the advantages of price-based controls are not so clear as this picture suggests. First of all, the efficiency incentives to companies may only apply early in a price control period. Towards the end, when re-setting is being considered, companies may have an incentive to conceal or postpone price-cutting initiatives so as to encourage the regulator to impose a lenient X term in the next round (Chapter 11). Second, price controls do not in practice minimize information requirements. Indeed the price regulator needs to know just as much information to re-set X values as does the profit regulator. This is simply because the regulator cannot set an X term without knowing the cost structure of the regulated companies. This in turn means he or she must make an estimate of the capital expenditure necessary or likely in the next period and must take an explicit view of the rate of return 'allowable' on that investment. More generally the regulator needs to take an overall view of rates of return. This operates financially and politically. Financially, a rate of return that is too low in relation to other investments will mean that it will be difficult to attract sufficient funds into the industry and may threaten the viability of the companies involved. Politically, a rate of return that is unnecessarily high will lead to strong pressure for downward movement on prices. While it is probably true, therefore, that price regulation is preferable to rate of return, it cannot avoid the need to take an explicit view of an acceptable rate of return for the industry. This means that the regulator needs to delve deeply into the cost structure of the industry, and runs the risk that his or her limited access to information may be a source of bargaining advantage on the part of the regulated companies.

REGULATION OF THE CONTINUING MONOPOLIES – PRICES

When the ESI was privatized it was clear that the two transportation functions – transmission and distribution – had natural monopoly characteristics and would need to be regulated monopolies for the foreseeable future.[13] Price controls were therefore seen as primarily falling on these two functions. Because there would already be competition in generation – and this competition was expected to increase sharply – no specific regulation, and certainly no price regulation, was intended in this part of the industry. Supply was seen as in need of temporary price regulation. Thus there was an overall price supply control from 1990 to 1994 (with a separate control for franchise customers) but since 1994 there has only been a franchise supply price control.[14]

The absence of a price control in generation immediately seemed to offer the prospect of an industry already substantially de-regulated, because for consumers, generation accounts for 52 per cent of the retail price.[15] By contrast, transmission and supply controls were relatively unimportant because they collectively account for only 12 per cent of final consumer charges. Distribution was likely to be the most important directly regulated area because charges for the use of the local distribution system account for some 26 per cent of retail prices. The final component of price in England and Wales is the fossil fuel levy. While the mechanics of the levy are fixed each year by OFFER, its existence is the result of a Government policy decision to pay for the excess costs of the nuclear industry (and to a minor degree for renewables) via a tax on all consumers. This levy was essentially fixed, at first for the eight years 1990 to 1998 by Government during the privatization process, and it has constituted some 10 per cent of consumers' bills. It is another example of the lack of agreement between Government and OFFER, which has publicly expressed its dislike of this form of taxation of consumers (see Chapter 7).

The background to the price controls instituted in 1990 was that the Government had insisted, through raising the required rate of return on assets in the ESI, that electricity prices should rise sharply – by 15 per cent – in the 1987 to 1989 period (see Chapter 9). The price controls that were then enforced from Vesting at April 1990 were the work not of the regulator but again of the Government, in the process of negotiating the details of the privatization. Any criticism that the initial price controls for the companies were insufficiently tough (especially in the light of the price increases between 1987 and 1989) is therefore properly aimed at Government not OFFER. The timetable in England and Wales was to be that the transmission price control was to last three years, the supply control four years and the distribution control five years. This first cycle has now been completed, and new price controls are in place for all three functions, and to last until 1997, 1998 or 2000.

Transmission and Supply

For England and Wales, the initial transmission control was RPI–0, and the review in 1992/93 determined a shift to RPI–3 for the period 1993 to 1998.[16] Because transmission charges are only 5 per cent of domestic consumers' bills, this change was going to have limited effect, and the review aroused little public interest.

The re-setting of the supply price control for England and Wales was implemented in April 1994. The review which preceded it was more complex than for transmission.[17] The X factor was tightened from 0 to –2 and there were two other important changes. First, the RECs had previously been subject to an overall supply price control, including the above 1 MW market which had supposedly been fully competitive since 1990. All price controls were now abolished in the supply market above 100 kW. Second, £25 m was to be collected annually (approximately £1 per customer per year) to be used for energy efficiency investments within the ESI (see Chapter 9). Nevertheless the impact of the supply price control review was only slightly greater than for the transmission reviews, given that the new control applied to only 7 per cent of retail prices.

Distribution

The transmission and supply price control reviews were therefore more or less unremarkable – mainly because they had so limited an impact on final prices, and in any case introduced only limited changes to the X terms. Much more important, at all levels, were the re-setting of the distribution price control (England and Wales) and distribution and supply controls (Scotland) timed for April 1995. At one level, these were bound to be more politically prominent because distribution accounts for about a quarter of final bills. But there was a further major reason why the distribution price control reviews were likely to prove significant. At Vesting the distribution price controls had, for 11 of the 12 RECs, taken the form of RPI *plus* X rather than RPI minus X. The formula varied between RECs but all except London were allowed to charge up to 2.5 per cent more, in real terms, for distribution services, each year and over five years.[19] This was supposedly to allow RECs to carry out large and necessary new investments. But what these formulae – instituted, it should again be stressed, by the Government rather than OFFER – appeared to ignore was that RECs had scope for cost-cutting and were in a position of statutory monopoly in their distribution businesses. In practice RECs' profits grew steadily every year, and distribution has always accounted for over 85 per cent of their total profits (Chapter 9). From as early as 1992, it started to become clear that the virtually risk-free distribution business was earning unexpectedly high rates of return. This led to some pressure to open the new price control earlier than 1995, but in the interests of stability of the regulatory framework OFFER resisted such attempts.

The English review was a long-drawn process. OFFER's original price proposals seemed significantly tougher on the companies than the earlier reviews of transmission and supply. The proposals were for a one-off cut in charges, varying between 11 per cent and 17 per cent depending on the REC, to be implemented at April 1995.[20] Initially it seemed that these would, from a consumer perspective, simply act as a partial offset to the raising of VAT from 8 per cent to 17.5 per cent at the same moment, but the political failure to implement this second stage of VAT meant that consumers had a direct price cut of up to 3 per cent or 4 per cent from April 1995. The X term from 1996 to 2000 was to become –2. OFFER calculated that if the new controls were expressed in terms of an average annual X value over the whole period of the new control, this would vary between –5.5 per cent and –7.5 per cent depending on the REC involved. This was on the face of it a major tightening.

It was not, however, the end of the story. In late 1994 Trafalgar House announced a hostile bid for Northern Electric REC. This was an unexpectedly early development, given that the Government's 'golden share' in the RECs, preventing any significant restructuring of ownership, was not due to expire until the end of March 1995. Northern's management put together a package of defence measures against the bid that promised its shareholders over £500 m in benefits. OFFER did not react immediately because Government was in the process of floating the remaining 40 per cent of National Power's and PowerGen's shares still in public hands. However, immediately after the flotation was complete, OFFER announced in March 1995 that the distribution price control was to be re-opened, arguing that the Northern defence showed that the company was stronger financially than it had revealed during the distribution review process.

The OFFER announcement was a shock to the industry. It had hardly occurred to anyone that the regulator was constitutionally entitled to re-open the distribution price review just days before his proposals were due to go into effect. The DGES ruled that the one-off price cuts for April 1995 would stand, but that the remaining period covered by the Review (April 1996 to March 2000) would be re-considered over the next three months.

The simultaneous review of distribution and supply price controls in Scotland was to play a significant role in the result of this re-opened review.[21] Hydro-Electric believed that OFFER's proposals for tighter price controls were too stringent, arguing that significant new investments were needed. It therefore refused to accept the new price controls and the consequence was that OFFER had to make an MMC reference in November 1994 to adjudicate on the issue,[22] the first direct involvement of the MMC in the affairs of the privatized ESI.

The MMC report of June 1995 made marginal changes to the price controls for Hydro-Electric, but more interesting was its reasoning.[23] It argued that OFFER had over-valued Hydro's assets, and that redundancy costs – which OFFER had accepted as over and above 'normal' costs – should be absorbed by the company. In the Hydro case, these two factors made very little difference to the X-factor results. However, OFFER chose to adopt the MMC's principles in its re-opened review of the distribution control for England and Wales. In the English case the new procedures had the practical effect of making the price controls significantly tighter.

In the revised distribution review, OFFER argued that application of the MMC's asset valuation procedure would reduce the value of the RECs' assets.[24] At the 7 per cent rate of return that the DGES regarded as legitimate this would sufficiently reduce total returns as to allow a further one-off cut in distribution charges in April 1996, ranging from 10 per cent to 13 per cent depending on the REC involved. OFFER then argued that the ruling out of redundancy costs would permit the subsequent annual value of X from April 1997 to be changed from –2 to –3. The overall result was that REC distribution charges would be 31 per cent lower in real terms by 2001 than they were in March 1995.

What is interesting about these changes is their highly technical nature.[25] OFFER chose not to pursue directly its earlier argument that Northern's defence bid had shown the RECs to be much stronger financially than they had revealed, so that an across-the-board tightening should take place. Instead the DGES simply adopted the difference in accounting practice advocated by the MMC and applied it directly to the RECs. This seemed more of an admission of technical error on his part than a policy shift towards a tougher regime.[26] The overwhelming advantage to OFFER in adopting this technical approach was that it virtually guaranteed that the RECs would have to accept the new formulation. OFFER's adoption of the procedures just enunciated by the RECs' potential 'court of appeal' (the MMC) meant that the RECs would have virtually no chance of successfully challenging the new formula, no matter how unpalatable.

As soon as this more stringent distribution price control was finally settled, there followed a whole spate of bids for the RECs by companies anxious to acquire the low-risk cash flow that the new controls virtually guaranteed to the end of the century (see Chapter 10). So while consumers were clearly gaining extra benefit from the revised controls, the markets clearly still judged the RECs as a highly attractive investment.

REGULATION OF THE CONTINUING MONOPOLIES – STANDARDS OF SERVICE

In addition to controlling elements of the retail price, it was in the area of standards of service that direct protection of franchise customers was expected to operate.[27] Given that the RECs retained monopolies in distribution indefinitely, as well as in supplying small customers until 1998, it was also expected that some weak version of 'yardstick' competition could operate in this area, with RECs vying to ensure they finished above average on issues such as offering and meeting appointment times, restoring supplies, obtaining meter readings, etc. The difficulty, in terms of consumers noticing any difference, was that the standards of service inherited from the publicly owned industry were already rather high.

In July 1991 OFFER introduced two kinds of service Standards (Chapter 9). One concerned 'Guaranteed' Standards, where customers would be entitled to financial compensation if the REC failed to meet the required level. The second was 'Overall' Standards where direct compensation could not practically be paid but an overall standard could be applied. These Standards were tightened in 1993 and in 1994 it was announced that further improvements would be required in the future. The post-privatization period has seen further modest but useful improvements on an already good record. (Chapter 9 has a much fuller discussion of the standards of service in terms of results.)

OFFER also specifies Codes of Practice for the RECs in relation to a number of important customer issues, including payment of bills, services to the elderly and disabled, the giving of objective energy efficiency advice, and the handling of complaints. In this latter area, the level of complaints against RECs has fallen significantly in each of the last two years.

The DGES has also, like his counterpart at OFGAS, taken the view that consumer disconnections should effectively be ended, but that RECs be allowed to implement ways to ensure that bills are paid in full. There has been a large growth of pre-payment meters, largely motivated on the part of RECs by the desire to ensure payment of bills, and 1 in 8 domestic customers now use this type of meter. The number of annual disconnections has fallen rapidly since March 1990 (when it ran at an annual rate of 80,000): in 1994 the annual rate had dropped to the virtually negligible annual level of 1200 (Chapter 9). This is undoubtedly an advance, though it does not of course ensure that poorer consumers are able to afford electricity any more than before.

REGULATION OF POTENTIALLY COMPETITIVE MARKETS – GENERATION

Whatever the merits of the RPI–X formula and its implementation, OFFER was at least, in the cases of transmission, distribution and (the franchise part of) supply, on the traditional regulatory ground of controlling monopoly power. In generation, which started with a partially competitive structure and was expected to become strongly competitive, there were to be no price controls. Instead OFFER's job was the relatively uncharted one of encouraging the rapid emergence of more competition.

The earliest area of regulatory interest in generation was the operation of the Pool. (While the Pool brings together all the main functions of the ESI, it is essentially a one-sided – generator-dominated – market. In this important sense, regulation of the Pool amounts to regulation of generation.) As early as 1991 OFFER published a Pool Price Inquiry[28] which recognized that generator market power might eventually lead to a MMC reference. It also found PowerGen to be manipulating pool prices by 're-declaring' plant as unavailable for generation at the last minute. New licence conditions forbade this practice.

By late 1992, the DGES was explicitly referring to the problematic nature of the market power of National Power and PowerGen in the Pool.[29] The basic problem was that the Pool price for every half hour was set by the last (marginal) plant to have its price bid accepted. Such plant was by definition not being run by generators on 'base-load' – round the clock, round the year, and in practice all the other generators were running their plants in base-load mode. Effectively therefore it was virtually always National Power or PowerGen plant that fixed Pool prices, and there was no immediate possibility of competition in this area. In 1993 the DGES explicitly threatened that if National Power and PowerGen did not comply with his plans to speed up the evolution of competition, he would refer them to the MMC.[30]

By February 1994 the two big generators complied. OFFER's terms were that the two companies would have to use 'all reasonable endeavours' to sell 6000 MW of coal- and oil-fired plant (4000 MW from National Power and 2000 MW from PowerGen) by the end of 1995.[31] The hope was that such plant would have a high chance of being operated in mid-merit or peak mode and thus offer direct competition in Pool price-setting to the incumbents. Secondly, and in many ways surprisingly, the two generators were to behave, in their price bids to the Pool, in such a way as to ensure that the average annual Pool purchase price would not exceed 2.4p/kWh (time-weighted) or 2.55p/kWh (demand-weighted) in the two years from April 1994. The reasoning behind the imposition of this Pool price ceiling was that the generators had enough market power to raise prices above competitive levels.

This Pool price ceiling made it completely clear that the Pool was not a very effective market, not even (now that it was to be price-controlled) a spot market. While the agreement was for a ceiling on prices it was in practice a price-fixing device (it would hardly be in the generators' interests to undershoot). Its impact on the two large private generators was limited by the fact that they had already sold most of their output for the two years in question on the contract market, which was unaffected by the new control. Worst affected, though not implicated in the original 'crime' was Nuclear Electric, which had less contractual cover and whose sales of electricity were much more exposed to Pool price variations.

The implications of OFFER's actions were profound. As in the case of gas regulation, a single individual was now sufficiently powerful as to shape the evolving structure of the generating industry, despite having no formal power in the Electricity Act to require the generators to act as they have agreed. The DGES's actions were perhaps a measure of his frustration at the very limited degree of real competition that was evolving in the generating market. That the generators agreed to his plans presumably reflected their fear that the MMC would argue that their monopoly powers were indeed extensive and potentially subject to abuse. This could in principle lead to radical recommendations – including a break-up of the two big generators – and OFFER's proposals, however unpalatable, were preferable to taking that risk. More than any other regulatory act to date, the Pool price cap and plant divestment requirement established that while regulation might begin to fade away at some future moment, it was certainly becoming more powerful, and even arbitrary, in the short term. The actions of OFFER were now not only seen to be crucial in the natural monopoly parts of the industry, but in the competitive parts as well. Here was a clear case in which the idea that regulation would get lighter as competition emerged was directly contradicted – more competition implied a need for more regulation, not less.

OFFER was active in other areas of generation regulation, though with nothing like the impact of the Pool price ceiling. In 1992/93 the DGES carried out a Review of Economic Purchasing, designed to check that RECs were not unduly favouring their own (part-owned) CCGTs at the expense of coal-based contracts offered by the big private generators. In a two-stage process, OFFER found that the RECs had behaved appropriately.[32] It might be the case that the *costs* of coal-fired power were lower than from CCGTs, but the RECs had to face contract *prices*, and the finding was that the prices they were paying for gas-based power were not unduly expensive in relation to the price of coal-based electricity. OFFER rather turned the focus back to the generators and announced that the generators' margins and costs might need closer regulatory attention.

In 1993/94 other Pool reforms were considered but largely – for the moment – rejected. There were three main areas: demand-side bidding, trading outside the Pool, and giving generators their Pool bid price rather than paying the marginal price to all generators.[33] In the first two of these cases OFFER expressed sympathy with the objectives. Limited experiments with demand-side bidding had yet to show clear results, and further action might be premature. In the case of trading outside the Pool, the reasoning was that too little was known about potential benefits, but that significant costs and the risks associated with a 'thinner' market meant that the case was not yet proven. In relation to paying generators their bid price rather than the marginal price – a potentially radical reform – there was no enthusiasm from OFFER. Risks, especially to smaller generators, might rise without a strong expectation of price falls. Paying bid prices therefore seems to have been ruled out of further consideration at least for some time to come.

REGULATION OF POTENTIALLY COMPETITIVE MARKETS – SUPPLY

The progressive lowering of the REC supply franchise limit from 1 MW in 1990 to 100 kW in 1994 was accompanied by detailed but strategically unimportant regulatory activity. The case of the intended complete abandonment of supply franchises in 1998 will clearly be different. The issues here are very large (and documented at greater length elsewhere in this book, especially Chapters 3, 10 and 11). There is no real precedent elsewhere in the world for throwing open choice of supplier to all domestic customers.

The specifically regulatory issues cannot yet be discussed in any detail because the regulatory requirements for this final liberalization are yet unknown.[34] But it is clear that the opening up of the domestic supply market will give rise to major regulatory needs. Among the most obvious source of these needs is that there will no longer be any company in a given area with a responsibility to ensure that all customers with 'reasonable demands' can obtain a supply. It is therefore virtually certain that OFFER will need to intervene so as to achieve a regime in which particular companies (probably incumbent RECs) are given some residual connection and supply obligation, and it will need to consider whether or not competitor companies will be required to finance this security requirement. A second area, already highly visible, is the enormously complex trading and settlement arrangements that will need to be put in place for competition to work properly: OFFER will certainly need to adjudicate here. A third area is that competition in the domestic market will almost certainly tend to work in favour of larger domestic consumers able to pay their bills promptly and through the banking system. The dismantling of cross-subsidy will also, therefore, lead to very different terms of supply and tariff levels between ostensibly similar customers, and it is easy to see how this may lead to dispute about due or undue discrimination. Finally, the new competitive market is to work through a system of 'profiling' (imputing the time profile of consumers' use of electricity from their total consumption and a 'standard' pattern of demand) rather than installing new meters, and the possibilities for consumers to dispute their allocation to a particular profile seems in principle almost endless.

The Trade and Industry Select Committee was sufficiently concerned about the slowness and incompleteness of preparations for 1998 to propose eight specific preparatory actions to be undertaken by OFFER (which does have a Task Force to prepare for domestic competition).[35] The introduction of this final liberalization certainly illustrates graphically how the introduction of more competition in a market for a basic service like electricity will be the occasion for a large volume of new regulation rather than its reduction. To equate domestic supply competition with deregulation is clearly to miss the point.

Environmental Regulation

The treatment given to environmental regulation is much briefer than to economic regulation, because many of the issues involved take us beyond the scope of assessing electricity privatization and into wider questions of environmental policy. Nevertheless, as the introduction to this chapter argues, environmental regulation has been a potentially important *economic* issue for the electricity supply industry, both before privatization, but with growing force ever since. The purpose here is to look at the interactions between environmental policy and the economic decisions and results for the privatized industry. We are interested both in the effect that environmental regulation has had on the privatized ESI, and also in the impact the privatized ESI has had on the environment.

SCOPE AND FRAMEWORK

Environmental issues are clearly multi-faceted, and the scope ranges from the purely local to the global. For the present purpose we limit ourselves to the major issues that have had a potentially large impact on ESI behaviour. This means that we concentrate broadly on air pollution issues. There are three that identifiably have significance to ESI strategy: climate change and carbon dioxide emissions; the control of sulphur and nitrogen emissions from large combustion plant determined at European level; and requirements under the integrated pollution control (IPC) framework for combustion plant to obtain authorizations for discharges at individual power stations.

An important issue is the tension in control philosophies between macro or national level requirements, as in the case of carbon dioxide and sulphur/nitrogen emissions negotiated internationally, and the plant-specific requirements that have been important in the IPC authorizations for discharges. We do not treat other issues, such as environmental impact assessments for new plant and local planning requirements, which have not, as it turns out, been significant issues in economic terms.

The period since the late 1980s has been one of growing activity in environmental regulation, and the ESI – whether reformed or not – would clearly have had to pay much closer attention to environmental questions in its strategic planning over the last few years. This is particularly so for the generating side of the industry where most of the larger and non-local environmental issues have arisen.

One important way in which environmental regulation differs from economic regulation is in the role and influence of much wider political communities than the nation-state. All of the three areas covered in this section have been subject to international negotiation. At the widest level, the Framework Convention on Climate Change – with associated carbon dioxide emission reduction targets – is negotiated globally (eg at the Rio

Earth Summit of 1992 and the Berlin meeting of 1995). At the next level, overall targets for sulphur and nitrogen emissions have been most recently negotiated through the UN's Economic Commission for Europe (ECE), though in the late 1980s this had been done through the European Commission. Finally, the basis for discharge authorizations at individual plants has been settled via the European Union.

These processes do not mainly consist of international impositions on the nation-state, because in all three cases national Governments retain some control over the determination of national targets. Nevertheless in these very important areas of environmental policy, it is only possible to determine detailed outcomes within a context of international negotiation. The need to incorporate Europe-wide rules into national legislation is now a significant issue.

The role of national policy in the context of the three issues covered here is mostly the need for international agreements and targets to be translated into national legislation, and into more concrete and specific targets at national and sub-national level. Thus carbon and sulphur emission limits are set internationally at an aggregate level for the UK as a whole: national policy must then determine targets and requirements for the specific sectors and even in some cases companies. Policy responsibilities in the UK lie with the Department of the Environment, and the legislative framework is largely set by the Environmental Protection Act (EPA) of 1990 and The Environment Act of 1995.

The origin of current UK environmental control strategy is in the idea, first officially mooted in 1976, of integrated pollution control.[36] This arises from the observation that control of emissions in one medium can simply add to the burden of pollution in another medium, so that it is necessary to create a unified inspectorate with responsibility for minimizing environmental harm across all three media of air, land and water. Eventually, in 1987, Her Majesty's Inspectorate of Pollution (HMIP) was created, with a view to implementing IPC systems. However the implementation of IPC was not embodied legislatively until the passage of the 1990 Environmental Protection Act. A principal purpose of the 1990 Act was to use the IPC idea in the fulfilment of international environmental control obligations, in particular the 1984 European Directive on Air Pollution from Industrial Plants[37] (which requires individual plant standards for both new and old plant) and the 1988 Large Combustion Plant Directive,[38] which sets national targets for the reduction of sulphur and nitrogen emissions to the year 2003. In 1995 a new Environment Act led to further integration in environmental policy by amalgamating HMIP functions with those of waste regulation authorities and the National Rivers Authority in a single new Environment Agency, which came into existence in August 1995.[39] At the level of the European Union, an Integrated Pollution Prevention and Control Directive was agreed in 1995, stressing the importance of integrated approaches in somewhat similar fashion to the UK's legislation. It differs from the UK's own IPC approaches in emphasizing energy efficiency, rational resource use and prevention to a greater extent,[40] though the implications of this Directive for the British ESI are not yet clear.

Integrated pollution control is underpinned by two main concepts. These are the ideas of BATNEEC (Best Available Technology Not Entailing Excessive Costs)[41] and BPEO (Best Practicable Environmental Option).[42] BATNEEC is inevitably an imprecise notion and open to significant regulatory discretion. It also explicitly applies differentially according to whether the process under review is new or existing.[43] The criterion is more stringently applied to new processes, where the impact of standards on profitability is to be disregarded, and processes may still be refused authorization if best available technology still leaves serious environmental harm. For existing processes, the intent is to upgrade plant to new plant standards, but a significant time may be allowed to elapse before this criterion is fully applied. BPEO applies where damage may be done across more than one medium, and establishes minimum environmental damage across all media, for the long term as well as the immediate future.

CARBON DIOXIDE EMISSIONS

Emissions of carbon dioxide only became a practical environmental issue following the Earth Summit in Rio de Janeiro in 1992, although the European Community had developed prior plans (eventually rejected by the Member States) for a combined carbon/energy tax to help counter carbon dioxide emissions. Among the agreements made at Rio in the Framework Convention on Climate Change was a commitment to the publication of a national plan for carbon dioxide emissions for the year 2000. The UK's was for a return by the year 2000 to the levels of carbon dioxide emissions in the year 1990 ('stabilisation'), and the British Government published plan suggests that emissions under 'business as usual' would have risen from the 1990 level of 158 million tonnes to 168 million tonnes in 2000.[44] If these forecasts were accurate they implied a need to find 10 million tonnes of carbon reductions over the rest of the 1990s. The intention was to achieve these savings by voluntary action, much of it to be coordinated by the Energy Saving Trust, a non-statutory agency with a budget that was originally to be largely drawn from levies on gas and electricity consumers.

The proposed savings were widely spread across the household, services and business sectors, with the electricity sector expected to achieve small savings by the substitution of gas for coal in generation. The combined effect of lower carbon content in gas than in coal, and the higher conversion efficiencies of gas in the gas-fired combined cycle means that the use of gas for electricity generation leads to only half the carbon emissions per kilowatt hour of electricity generated when, as is currently the case, new gas-fired generation directly substitutes for coal-firing. The National Plan did not suggest that this substitution of gas for coal would involve any change in investment policy for the generators, and in the event the substitution of gas for coal has proceeded much more quickly than was foreseen even in early 1994 (see Chapter 6). Such has been the speed of this fuel substitution that some 17 million tonnes of carbon will be saved in 2000 compared to 1990 solely as a result of rapid substitution of gas for coal.[45] In the carbon dioxide area, then, environmental targets never looked likely to force any policy change among the generators; conversely, rapid substitution of gas for coal, driven by commercial pressures in the new privatized electricity system, seems to be leading to a comfortable over-fulfilment of carbon-reduction targets.

Overall Sulphur and Nitrogen Limits

The issue of acid deposition and acid rain has been on the policy agenda in Europe since the early 1980s, and there were long negotiations over a proposed European Union Large Combustion Plant Directive from the mid-1980s. Early British approaches were sceptical of the need for large sulphur or nitrogen emission reductions because of supposed scientific uncertainties about the level of damage caused by such emissions. However the UK changed its position during the course of the negotiations and the Large Combustion Plant Directive (LCPD) of 1988 committed the UK to reduce national sulphur emissions (compared to 1980 levels) by 20 per cent in 1993, 40 per cent by 1998 and 60 per cent by 2003.[46] These were less demanding targets than those agreed by other

North European states: Germany, France, Belgium and the Netherlands all needed to make reductions by 40 per cent, 60 per cent and 70 per cent by the same dates.

The lighter treatment won by Britain was influenced by the appearance that Britain would need to invest substantially in the expensive retrofit technique of flue gas desulphurization (FGD) if even the 60 per cent target were to be achieved. Adding FGD to existing coal-fired plant is estimated to raise generating costs by as much as 0.55p/kWh, or about 25 per cent.[47] Originally it seemed probable that up to 12 GW of plant would have to be fitted with FGD to comply with the Directive, but by the time of publication of the relevant British national plan, this had fallen to 8 GW, and the final total now looks likely to be as low as 6 GW (4 GW by National Power and 2 GW by PowerGen). The implementation of the aggregate target for sulphur reductions has mainly focused on the ESI, and both National Power and PowerGen have been required to meet specified reductions in sulphur emissions for each target date. These company targets take the form of 'bubbles'[48] (aggregate reduction levels but with no requirements as to the means of achievement).

Both generators seem able to meet their sulphur reduction targets quite comfortably without in any way altering the investment strategy that environmentally unconstrained decisions would have produced. In relation to sulphur control, the advantages of gas over coal are even greater than in the carbon case because gas is essentially sulphur-free. This means that the generators will probably be able, such is the volume of their CCGT investment and reduction in coal use, to meet their 'bubbles' without much use of FGD installed (PowerGen for instance does not always base-load its FGD-fitted station at Ratcliffe).

The LCPD contained provision for re-negotiation, but this process was essentially superseded by a wider negotiation – across the whole of Europe – by the Economic Commission for Europe's Sulphur Protocol, which has specified further reduction in sulphur emissions for the years 2005 and 2010.[49] In the UK case these reductions are of 70 per cent and 80 per cent respectively. Again, present indications are that the switch to gas from coal in the power sector will probably be adequate, together with expected increases in gas use elsewhere in the economy, for these targets to be achieved without any real interference in the commercial logic of investment decisions.

DISCHARGES FROM INDIVIDUAL PLANTS

The LCPD also contained much stricter rules for emissions from new plants. These effectively mandate the fitting of FGD to any conventional coal-burning plant. The effect over the last decade of this need for FGD on new plant has been a significant disincentive to generating investment based on coal (even though the CEGB was contemplating such investment up to 1988 at least). This undoubtedly played a role in the attractiveness of gas as a new investment option following privatization: at the same time that gas was becoming cheaper and gas-using technology more efficient, competition from coal was being hampered by the much higher private costs that the LCPD imposed on conventional ways of burning coal.

Before the LCPD came into operation, the European Commission issued a directive in 1984 on Air Pollution from Industrial Plants. This required all existing and new large combustion facilities to obtain authorization from national environmental authorities. The British Environmental Protection Act of 1990 allowed this directive to be translated into national action. The most important issue here is the provision to adapt existing plants, over time, so that they eventually embody best available technology, though the BATNEEC criterion (especially avoidance of 'excessive cost') is also relevant.

Interpretation of the Directive is a national matter and HMIP originally intended that 'all existing large combustion plant should be upgraded to the levels for new plant whenever the opportunity arises and by April 2001 at the latest'.[50] If this had meant that all coal-fired plants would have needed FGD by 2001 this would have profound implications for generator strategy. Given the expense of retrofitting FGD, it would have implied a need to confine coal use to the 6 GW of plant already fitted with FGD. In the face of this apparent threat, the generators undertook a long process of negotiations with HMIP on the precise implications of the Directive for plant standards in 2001 and beyond.The grounds for such lengthy negotiation are inherent in the idea of BATNEEC: the regulator can stress the need for 'best available technology' while the generators can emphasize the 'net entailing excessive cost' provision.

By March 1996 the negotiations were complete and HMIP issued new discharge authorizations.[51] The Inspectorate applied no requirement for the fitting of specific technologies like FGD to individual stations, but instead introduced new and more stringent emission limits for individuals as well as even more stringent 'bubbles' to National Power and PowerGen. These are intended to ensure that the new environmental impact of the new standards will be *equivalent* to the effects of fitting FGD to all coal- and oil-fired stations in use in 1991. Compared to actual discharges in 1991, sulphur dioxide emissions wil need to fall by 79% by 2001 and 85% by 2005. In the absence of new and very expensive FGD investment, this means either that coal-fired stations will need to run at very low loads, or that the owners of the stations will need to convert them to gas-firing.

In the case of National Power and PowerGen, these new and quite stringent limits may have relatively little impact, as the companies probably plan to run their coal-fired stations at relatively low load factors after the turn of the century. Nevertheless, at the margin they introduce some further incentives to accelerated gas use by conversion of coal-fired stations to the use of gas. A larger incentive to more rapid conversion to gas may emerge from the divestment of 6000 MW of coal-fired plant by National Power and PowerGen. The new company and companies that will own this 6000 MW may be expected to want to run the plant at high load. If they are given a 'bubble' that is equivalently restrictive to the bubbles given to the incumbent generators this would imply a need to run the divested stations at less than 25% of capacity even if low-sulphur coal was burned.[52] For the new owners of divested plants, who will very likely want high outputs from them, the new authorizations seem to offer a strong incentive to converting them from coal-firing to the use of gas. As in all other relevant areas of environmental policy, rules on individual plant standards and company bubbles tend to push towards gas use rather than coal.

CONCLUSIONS

Economic regulation in England and Wales has in one sense been ambitious and wide-ranging – encompassing the orchestration of competition as well as the control of monopoly – but at another level very narrowly focused on efficiency and competition issues. OFFER has been the dominant agency of regulation, and has paid little explicit attention even to some wider issues within its remit, for instance energy efficiency, the environment and R&D, let alone some of the more political issues of equity treated in Chapter 11. The Government believed initially that regulation would start light and get lighter, because the advance of competition would directly allow the retreat of regulation. Competition has advanced at least somewhat, but organizing it has required more regulation rather than less, and there is no sign yet of any trade-off between competition and regulation. Controls, never envisaged in the original regulatory system, were put on the Pool prices in 1994, and the introduction of retail competition in the domestic market will almost certainly lead to a host of new regulatory activity, both in the design of the new system and in its continuing operation.

In the area of monopoly regulation OFFER has been much criticized for the very high profits of the RECs accompanied by limited and slow falls in consumer prices (Chapter 9). The original regulatory settlement was undoubtedly generous to the RECs, and OFFER was unwilling to bring forward the distribution price review, which would have been the main vehicle for redistributing economic surplus between shareholders and consumers more quickly. When the distribution price review did take place, there was a classic case of the regulator's disadvantage in having less information than the regulated – only by good fortune did the regulator get the information needed to re-open the distribution price control and impose a more stringent settlement. Electricity is destined to remain a highly managed or regulated market for an indefinite future, and OFFER – for better or worse – will remain a powerful force in determining the shape and profitability of the industry as a whole.

In the almost totally separate sphere of environmental regulation, the interesting question has been whether environmental regulation has significantly modified, or made more expensive, the decisions taken by the privatized industry, or whether conversely, privatization has made much difference to the environment. The answer seems to be that environmental regulation has had a strictly limited effect on decision-making in the privatized industry, except to reinforce the private commercial logic of investing in gas-fired systems more or less exclusively, and possibly to speed up the rate of that investment given the potentially high cost of retrofitting coal-fired plant after the turn of the century. Looking at the question the other way round, privatization has made a real environmental impact, though almost entirely by the fortuitous route that the commercially cheapest investment strategy happened to be cleaner than the plant displaced by the new capacity. The rapid advance of gas, which is mainly a phenomenon of the new structure if not of privatization in the narrow sense, has made compliance with major environmental targets much easier than anticipated, and has led to lower emissions of sulphur and carbon than would otherwise be the case. Against this, the 'dash for gas' in power generation raises longer-term issues of supply security which are discussed in Chapter 13.

6

Effects on Demands for Fossil Fuels

Mike Parker

Introduction

Major changes took place in the use of fossil fuels for power generation during the first five years of electricity privatization (Table 6.1). The total use of fossil fuels has been falling since electricity privatization, reflecting in large measure the improved performance of nuclear plant within a broadly stable total fuelling requirement. Also, there has been a significant shift to natural gas away from coal, whose share of this market fell from 68 per cent in 1990 to 51 per cent in 1994. By the end of 1994, the capacity of gas-fired combined-cycle gas turbine plant (CCGT) already operating or committed was sufficient to make it almost certain that by 1998 the use of coal in power stations would be no more than about half the level in 1990. This major re-orientation of fuel supply can be understood only by looking at the situation prior to 1990.

The Coal Issue before Electricity Privatization

It is important to emphasize the symbiotic nature of the relationship between the electricity and coal industries prior to privatization. Over the period 1976/77 to 1989/90 (excluding the strike year of 1984/85), coal consumption of power stations averaged 83 m tonnes a year, representing some 73 per cent of total fuelling – and in 1989/90 was still 81 m tonnes,

Table 6.1 Fuel used in power generation: major power producers, 1990–1994 (mtoe)

	1990	*1991*	*1992*	*1993*	*1994*
Coal	82.6	82.0	77.0	64.1	60.7
Oil[1]	11.6	9.9	8.4	7.4	6.1
Gas	negl.	negl.	1.7	10.5	15.4
Total fossil fuels	*94.2*	*91.9*	*87.1*	*82.0*	*82.2*
Non fossil[2]	28.0	30.6	32.9	37.3	37.5
TOTAL	*122.2*	*122.5*	*120.0*	*119.3*	*119.7*

Source: author, based on Table 46, *DTI Digest of UK Energy Statistics 1995.*
1 Including orimulsion.
2 Nuclear, hydro, renewables and net imports over 'French Link'.

or 67 per cent of total fuelling. Over the same period, nearly all of the coal supplied to power stations in Britain had come from British Coal. British Coal's sales to power stations had averaged 81 m tonnes a year, representing some 72 per cent of its total sales – and in 1989/90 were still 76 m tonnes, equivalent to 80 per cent of British Coal's total sales. In the period leading up to electricity privatization, coal supplied by British Coal therefore still dominated fuel supply to the power stations and, in turn, British Coal was almost completely dependent upon the power-station market. It is this relationship between the coal and electricity industries which was to be fundamentally altered following electricity privatization. But the causal links are by no means straightforward, and they require an examination of Government policy towards the coal industry as much as ESI privatization as such.

The market position of UK coal had been relatively secure in the wake of the OPEC price increases in the 1970s, given that the only real flexibility available to the CEGB was to switch load between coal-fired and oil-fired stations, and that heavy fuel oil had become uneconomic following the increases in crude oil prices. The difficulties experienced in increasing the nuclear power contribution, and the absence of a large-scale international trade in steam coal until after the second oil 'price shock' of 1979/80, further reinforced the economic dependence of the power stations on UK coal. However, by the late 1980s, circumstances had changed considerably. The collapse of oil prices in 1986 radically altered perceptions about security of supply and the importance of indigenous energy and, by the late 1980s, the main potential competition to British Coal was seen as coming from internationally traded steam coal – increasing volumes of which were becoming available in Western Europe. International coal prices, combined with the relatively high value of sterling against the US dollar (due to strict UK monetary policy and high oil revenues), meant that in sterling terms these prices stayed broadly constant in 'money of the day'. Although rapidly rising productivity following the 'defeat' of the National Union of Mineworkers (NUM) strike of 1984/85 enabled British Coal to absorb UK inflation, the extent to which UK deep-mined operating costs exceeded international prices did not reduce. Over the four years prior to electricity privatization, UK deep-mined costs were some £0.50–0.90/GJ (£12–22 per tonne) higher than the potential delivered prices of imported coal, subject to the location of the power stations, while imported coal usually had the additional advantage of lower sulphur and chlorine contents than UK coal. Not surprisingly, in these circumstances, the electricity industry, and particularly the CEGB, greatly resented its dependence on British Coal, which it saw as the (unjustified) result of measures to protect British Coal and its employees from market forces. It was widely expected that electricity privatization would give the generators the freedom to reduce their use of high-cost, low-quality UK coal.

Mrs Thatcher's Conservative Government had much sympathy with such aspirations. Indeed there was a strong 'political agenda' in the Government's coal policy which would also lead in that direction. The Government's fundamental policy objectives for the coal industry emerged as early as 1979/80 and were held consistently thereafter, although the underlying policy remained largely covert until after the 'defeat' of the 1984/85 NUM strike. This policy had a strong political complexion and comprised two main elements: first, to break the power of the NUM 'to hold the country to ransom'; and, second, to subject the coal industry increasingly to market forces, in order to change it from the archetypal 'union-dominated nationalized monopoly, dependent on state funds', into a profitable business which could eventually be privatized, thereby demonstrating the greater merit of 'free enterprise' over 'socialist' nationalization. These two elements of policy were mutually consistent and reinforcing. They required reductions in the size of the industry and above all of its workforce, as far and as rapidly as practical considerations and political courage allowed. These policies were clearly and publicly articulated subsequently in the published memoirs of some of the principal actors: former Energy Secretaries Nigel Lawson and

Cecil Parkinson, and Margaret Thatcher herself. Writing about the 1984/85 NUM strike, Nigel Lawson said:

> Just as the victory in the Falklands war exorcised the humiliation of Suez, so the eventual defeat of the NUM etched in the public mind the end of militant trade unionism which had wrecked the economy and twice played a major part in driving elected governments from office.[1]

On his pledge at the Conservative Party Conference in 1988 to achieve the 'ultimate privatization' of coal, Cecil Parkinson wrote:

> What was ultimate about the proposed privatization of coal was that it would mark the end of the political power of the National Union of Mineworkers and would make the coal industry what it should always have been, another important industry, no more and no less important than many others.[2]

He added, 'I have never understood the argument that Britain somehow owes a great debt to the mining industry. The industry was given a privileged position and it abused the privilege'.[3]

The views of Margaret Thatcher were, if anything, even more strongly expressed. She said that, 'by the 1970s the coal mining industry had come to symbolise everything that was wrong with Britain',[4] and she regarded the 'defeat' of the 1984/85 NUM strike as a great symbolic victory which established 'that Britain could not be made ungovernable by the Fascist Left'.[5]

In the light of these views from the top level of the British Cabinet, it would be difficult to exaggerate the importance of the Government's political agenda on the coal industry in influencing the outcome of electricity privatization, British Coal's main market.

A further downward influence on the use of power-station coal also developed in the years before electricity privatization. It was the environmental problem. The adoption in 1988 of the EC's Large Combustion Plant Directive after about four years of negotiations, finally committed the UK to progressive reductions in sulphur dioxide (SO_2) emissions (requiring a 60 per cent reduction in power-station emissions by 2003, compared with 1980 levels). The view which was widely held at that time was that 12 GW of coal-fired plant retrofitted with flue gas desulphurization (FGD) plant would enable the use of moderate-sulphur UK coal to continue at broadly the same level. However, as electricity privatization approached, it became clear that the 12 GW FGD programme was not a firm commitment, given the large capital expenditure involved, and that the Government was content for the privatized generators to comply using also other cheaper methods, namely imported low-sulphur coal and gas-fired CCGTs. Eventually only 6 GW of FGD was to be provided in England and Wales. The rising interest in global warming from 1988 onwards also provided a further opportunity to Government to express its antipathy to the coal industry. Coal-fired power stations were often portrayed as the main danger to the planet. This is not to say that environmental concerns were decisive in increasing pressure to reduce UK coal use in power stations, but such concerns provided further weight and public justification for the underlying political agenda for coal.

By the late-1980s, in the run up to ESI privatization, the relationship of mutual dependence between the coal and electricity industries looked to be unsustainable. It appeared likely that, once the necessary port-handling facilities had been developed, most UK coal output would be uneconomic against imports. In addition, environmental concerns suggested a reduction in coal use. Above all, the Government was now in a position to move towards the fulfilment of its political objectives for coal as the scale of the defeat of the 1984/85 NUM strike had largely removed the ability of the NUM to confront the Government by threatening the security of the electricity system, particularly as a high level of power-station coal stocks was continued as a matter of policy after 1984/85.

The reasons for the decision to privatize electricity before coal are not clear-cut, but it was already clear that coal industry privatization would be difficult, both politically and practically, and certainly would not raise large sums for the Treasury as the ESI would do. It was also apparent that coal privatization would be even more difficult if it was carried out before the effects of subsequent electricity privatization were known. Whatever the inner reasoning, this order of doing things had certain advantages for the Government in relation to the coal issue. Firstly, because the ESI privatization process would itself be complex and protracted, this would allow more time for British Coal to close collieries and run down manpower in a more orderly fashion over a longer period. Secondly, any further restructuring of the coal industry prior to its own privatization in due course could be characterized as the result of commercial decisions by private electricity companies rather than action by the Government. Electricity privatization would unleash powerful forces to 'down-size' the coal industry 'by remote control'. This was recognized by Cecil Parkinson, who states in his memoirs that, 'it was obvious that the organization of the electricity industry, and the weakening of British Coal's monopoly as a coal supplier, would fundamentally change the shape of the industry'.[6]

During 1988 and 1989, the electricity industry made it clear that it wished to see a significant reduction in its dependence on British Coal, and for BC prices to move quickly towards parity with imported coal. In 1989/90 the BC pit-head price was £1.77/GJ, or around £2.00/GJ delivered – some 40 per cent higher than imported coal. Despite its evident sympathy with the aims of the new generators (National Power and PowerGen), the Government could not allow them complete freedom to impose on British Coal precipitate reductions in coal sales and prices. The success of the Government's coal policy required cautious implementation. It would appear that the Government acted on three principles. Firstly, coal contracts were required to provide an element of price stability in the period immediately following electricity privatization, particularly in the politically sensitive domestic market. For this to be done without prejudice to the finances of either the RECs or the generators, such contracts had to be 'back-to-backed' into the RECs' franchise markets under a framework of 'contracts for differences' so that the higher costs of UK coal could be passed through to final consumers; and coal prices needed to fall in real terms to increase the profitability of the electricity industry while maintaining franchise electricity prices broadly constant in real terms.

Secondly, the coal contracts needed to be sufficiently favourable to BC to avoid the Government having to deal with a 'second front' on coal until the power stations had been safely transferred to the private sector. Further, the form of the coal contracts would have to provide for sufficient volumes of coal sales to power stations to avoid large-scale colliery closures which could be attributed directly to ESI privatization, and coal prices which, although declining in real terms, would be compatible with BC's progress towards acceptable levels of profitability *without explicit subsidy*. The 'coal subsidy', which was the difference between BC prices and 'free market' prices based on parity with imports, should continue to be hidden in the coal price.

Thirdly, there was also the important issue of the duration of the coal contracts. Here, the reasoning appeared to be that the contracts had to be of sufficient duration and firmness to preclude any reopening until after the next general election (due by 1992) and to allow any subsequent radical contraction of the coal industry to be presented as the result of market forces operated by the privatized electricity industry rather than the direct effect of Government policy. But it was already becoming clear that further substantial contraction of the coal industry would be required *before* British Coal could be privatized. Thus, the duration of the new coal contracts could not be so long as to preclude the 'down-sizing' of the coal industry in time to privatize BC within the term of the following Parliament. Such considerations suggested contract duration of about three years.

The coal contracts which emerged were therefore a Government-imposed reconciliation of the policies towards the ESI and the coal industry. The initial contracts were for three years from April 1990 on a 'take-or-pay' basis with predetermined prices. Compared with 1989/90, where British Coal had planned sales of 75 m tonnes to National Power and PowerGen, there was a modest reduction in tonnage to 70m tonnes in 1990/91 and 1991/92, and to 65m tonnes in 1992/93, and prices were due to fall by the equivalent of about 5 per cent per annum in real terms.

These contracts failed to provide a basis for a long-term solution to the issue of the ESI's use of UK coal. The problem was merely shelved for a couple of years and, because delayed, was ultimately more difficult. It was recognized by all the parties that both the tonnage and the prices in the three-year contracts were significantly higher than a free negotiation would have produced, so that unless market conditions changed dramatically and unexpectedly during the currency of the contracts, it was highly likely that the generators would wish to diversify away from BC supplies and to reduce prices markedly once the contracts came to an end. Moreover, the creation of the duopoly of National Power and PowerGen in fact became a duopsony so far as the British coal industry was concerned. The coal industry had no significant alternative economic markets in the UK, or still less abroad, whereas the two generators had a wide choice of alternative fuels. There was thus a large imbalance of market power against British Coal, which was further exacerbated by Government policy to maintain power-station stocks at a high level as a further safeguard against the small possibility of effective strike action by the NUM. By October 1992, generators' coal stocks were some 33 m tonnes – equivalent to six months coal supply and more than 20 m tonnes higher than purely commercial considerations would suggest. This huge stock 'overhang' greatly increased the bargaining power of the generators by the time the coal contracts were due for renewal.

THE DASH FOR GAS

At the time of electricity privatization, it was widely expected that the main challenge to British Coal would come from increased coal imports by the two main generators. A figure of 30 m tonnes a year had been quoted both by the CEGB and its successors. Although new port facilities were developed to deal with this tonnage, no significant increase in coal imports took place. The main impact on coal came from a large and rapid programme of CCGTs which became known popularly as 'the dash for gas'.

While some gas-fired plant might have been built in any event, particularly as the use of gas in power generation was no longer regarded as undesirable on general energy policy grounds (recognized in the symbolic revocation in 1991 of the European Community directive discouraging such use), the size and speed of the 'dash for gas' in large measure arose from the policy of promoting competition in generation. In England and Wales, this entailed reducing the dominant market share of National Power and PowerGen, which at the time of privatization had been created with ownership of all of the existing fossil fuel capacity, which was predominantly coal-fired. It was politically unthinkable for the Government to contemplate breaking up National Power and PowerGen so soon after setting them up, or even at that stage requiring them to divest themselves of plant. The only way the market share of National Power and PowerGen could be reduced was through the construction of new generating plant not owned by either of these companies. The RECs saw CCGTs owned by the so-called 'independent power producers' as a way to achieve a measure of independence from the two main generators whose dominance they resented, even though this was not a way of reducing the market power of the two main generators since they controlled all the 'mid-merit' fossil-fuel power stations which determine half-hourly Pool prices most of the time.

The most cost-effective way to provide new generating capacity was to build CCGTs, because of their relatively low capital cost and short construction times. As the RECs, in association with oil companies and other independent parties, began their programme of CCGTs, the two main generators also began to build their own in order to protect their market share by ensuring that independents did not secure all the available gas supplies. Since CCGTs give rise to lower emissions of SO_2 and CO_2, they also provided a hedge against further environmental regulations which might restrict the use of coal-fired plant, that made up the bulk of the capacity of the two major generators.

The development of the CCGT programme was also influenced by the policy of regulation towards British Gas (BG). The promotion of competition by artificial props imposed by the regulatory bodies added to the rapid rundown of BG's share of the industrial and commercial contract market from 1992 to 1995. Even by 1993, this development effectively ended BG's previous status as monopsony buyer of gas 'at the beach'. Over the period 1989 to 1993, 36 new gas fields were contracted for, but only 9 of these went to BG, the other 27 fields being sold to 18 different organizations. Under the previous regime, new gas fields were bought into supply only to the extent that BG needed them to provide replacement for declining mature fields, or to meet incremental demand in BG's established monopoly gas markets. As competition developed, more gas became available because BG was no longer in a position to restrict total supply to the amount needed to satisfy forecast demand. The effect was a great deal of gas seeking additional secure outlets. Such markets could be made available with CCGTs, where the general rule was for 15-year gas supply contracts 'back-to-back' with contracts for sales of electricity to RECs who were almost invariably partners in the projects.

A subsequent investigation by the Trade and Industry Select Committee was to show that in most circumstances, the levelized lifetime generation costs of new CCGTs were higher than those at existing coal-fired stations, given the reasonable expectation of falling coal prices from BC. Price regulation allowed the RECs to 'pass through' into their prices any additional costs arising from contracts to purchase electricity from new CCGTs, provided that they could satisfy the regulator that they had met their economic purchasing obligation. However, they were able to do so because the regulator was concerned, not with costs, but with prices offered, and the RECs were able to show that the contracts they had signed for independent CCGTs were advantageous compared with the contract prices on offer from the two major generators, particularly as the former provided a hedge against future price rises caused by sulphur dioxide emission limits.

THE OCTOBER 1992 COAL CRISIS: ORIGINS AND RESOLUTION

The combined effect of all these factors which gave rise to the 'dash for gas' was that by March 1993 – the expiry date of the original three-year coal contracts put in place when the ESI was privatized – there were some 11 GW of CCGT plant operating, under construction or otherwise committed. And because these new stations were planned, through their contractual arrangements, to operate at high load-factors (typically 80 per cent), they would displace some 33 m tonnes of coal when fully operational by the mid 1990s. By the time the original coal contracts were due for renewal, British Coal therefore faced the near certainty that within a few years at least half of its 65 m tonnes of contract tonnage would be lost to gas. And, as we have noted above, there was the more immediate threat of the large 'overhang' of excess coal stocks at power stations, which the generators would wish to draw down as soon as possible, thereby displacing current sales.

On top of all this, the structure of the privatized electricity industry and its regulatory framework made it more difficult for BC to obtain favourable contracts after March 1993. There were three factors at work. Firstly, the two generators could not be persuaded to enter into new contracts on a 'take-or-pay' basis with pre-determined prices (the same basis as the original three-year contracts for 1990–93) unless these were once again 'back-to-back' with contracts with the 12 RECs for sales of coal-fired electricity in the franchise market. The price formula in that market included full 'pass through' of generating costs, thereby removing any commercial risk to the generators or RECs. But the franchise market was planned to diminish in 1994, when the franchise limit would reduce from 1 MW to 100 kW, and to terminate in 1998. And other forms of electricity generation, particularly CCGTs in which the RECs themselves had a financial stake, could be dedicated to the franchise market, thereby further limiting the scope for contracts for British Coal. Secondly, the committed CCGTs and nuclear power (with most of its output guaranteed a market by the non-fossil fuel obligation until 1998) would take nearly all the base load by the mid-1990s, leaving little scope for coal to supply the market at favourably high load factors. The third factor was that one of British Coal's best potential selling points was that it could, in principle, contract forward at predictable sterling prices, free of exchange rate risk and decoupled from the fluctuations of world oil, gas and coal prices. However, electricity regulation, which allowed the RECs to pass through increases in generating costs in the franchise market (subject to meeting the economic purchasing obligation), made it very difficult for BC to obtain a price stability premium. Also, the RECs were under an 'economic purchasing obligation' (overseen by the regulator) which would place limits on the price premium that could be paid for UK coal, at a time when the gap between UK prices and international prices showed no sign of diminishing, despite the high productivity growth in the British coal industry.

The accumulation of all the above factors meant that the use of British coal in British power stations, and the prices charged, were likely to be sharply reduced; and, for the coal industry, a difficult problem of further adjustment could become a cliff-edge crisis. BC took the reasonable view that it would be untenable to embark upon a further programme of large-scale colliery closures and redundancies until efforts had been made to secure the best possible contracts. Yet the size of the gap between what the generators considered reasonable in terms of volume and prices to apply after March 1993, and the parameters which BC could accept as providing a reasonable transition, was so great that between April and September 1992 the prospect of freely negotiated contracts receded.

The Government had been unwilling to press matters until after its re-election at the general election which took place in April 1992. Despite its concern for a resolution of the issue, the Government's difficulty in securing an acceptable outcome subsequently became all too apparent. The three-year contracts expiring in March 1993 had been the product of a Government 'fix' imposed at a time (1989/90) when all the parties involved were still under Government control as nationalized industries. But a further 'fix' was difficult to arrange three years later given the sheer complexity of the task of securing two contracts with the same terms between BC and National Power and PowerGen, but also 'back-to-back' with the 12 RECs. Given also that all the electricity undertakings were now private companies, that there was not supposed to be collusion between the generators or coordinated action by the RECs, and that the outcome had to be acceptable to both OFFER and the European Commission, the role of Government was constrained to say the least. The Government's position was also ideologically ambiguous. One of the objects of privatization was to remove the ESI from political influences, yet the Government could not remain neutral if its coal policy objectives were to be fulfilled in a politically acceptable way.

By September 1992 the Government, in its role as go-between, informed BC that new contracts could be agreed only if BC sales to the two main generators were reduced from 65 m tonnes in 1992/93 to 40 m tonnes in 1993/94 and 30 m tonnes between 1994/95 and 1997/98. BC was also told that prices would probably have to be reduced (at 1992/93 money values) from over £1.80/GJ in 1992/93 to around £1.50/GJ in 1993/94 and then to around £1.30/GJ by 1997/98. Extension of the new contracts beyond March 1998 was not considered in view of the planned termination of the RECs' franchises at that point.

After a long period of capacity reduction, BC was now faced with having to get rid of around half of its remaining deep-mined capacity, with its associated manpower, in about six months – a timetable made all the more rigid by Treasury rules on the time-limited availability of Government funding of the very generous redundancy costs considered necessary to secure manpower reductions without disruption. As a result, in October 1992 BC announced that 31 of it 50 deep mines would close in the near future, with the associated loss of 30,000 jobs. This announcement caused a national political storm known as the Coal Crisis of October 1992, which seriously threatened the Government's position in Parliament. The Government's first reaction was to claim that the British Coal closure announcement was the result of market forces, and that there was no case for Government intervention. As we have noted above, one of the elements of thinking at the time of ESI privatization was that this would enable the contraction of the coal industry to be induced 'by remote control' through action by the privatized electricity companies, rather than being seen as the direct result of Government action. This stance failed to satisfy the many critics of the Government, which was forced to concede that 21 of the 31 pits proposed for closure should continue in production for the time being, pending a full review in the context of the Government's energy policy.

The outcome of the Government's Coal Review was set out in the White Paper of March 1993, which was the first comprehensive statement of the Government's Energy Policy. On the fundamentals of the position, nothing was changed. The Government declined to intervene to restrict the introduction of new CCGT plant or the loading of such plant, to mitigate the likely sharp reduction in generators' coal stocks, or to restrict the use of nuclear plant or electricity imports over the interconnector with France. Nor were there to be any changes in the structure or regulation of the ESI. The only retreat made by the Government from its free market principles was the proposal to provide some subsidy to UK coal burned in power stations in displacement of imports. In fact, virtually no subsidy was paid and most of the capacity reprieved by the Coal Review was subsequently closed. By March 1993, coal contracts had been agreed under virtually identical terms to those proposals which had precipitated the Coal Crisis in 1992. The Government had defused the political crisis by playing for time and allowing the colliery closures to be spread over a few more months, by which time the public outcry had greatly subsided.

The new contracts provided for 'take-or-pay' sales of BC coal to the two major generators of 40 m tonnes in 1993/94 and 30 m tonnes in the subsequent four years; with pit-head prices (at 1992/93 money values) of £1.51/GJ in 1993/94 falling progressively to £1.33/GJ in 1997/98. These contracts were more the result of Government pressure and influence than of market forces. Such a combination of prices and volumes could not have been agreed in free negotiation between BC and the generators. The terms of the contracts owed much to the Government's policy towards the coal industry, in particular as a means of preparing the way towards BC's privatization, which required three preconditions to be satisfied. Firstly, the further radical contraction of deep-mined capacity and increases in productivity involving colliery closures and large-scale redundancies had to take place before BC privatization if there was to be a reasonable prospect of sale of BC's mining assets to the private sector. Secondly, the coal prices, although declining in real terms substantially, had to be significantly above parity with import prices in order to provide the new private owners of BC with sufficient profit without explicit subsidy to justify a politically acceptable purchase price. Finally, the take-or-pay volumes were needed to provide a period of stability after the privatization of BC, if only to allow the new owners to redeem their capital debt.

NEW PATTERNS OF ESI FOSSIL FUEL SUPPLY

The net effect of the interaction of market forces, electricity privatization, the improved performance of BC and Government policy in the coal sector has been that the demand for BC coal by the two major generators in England and Wales, which makes up the great majority of total power-station coal use was more than halved in the first five years of privatization, and prices fell in real terms by about a third (see Table 6.2).

What has happened so far will largely fix the pattern of fuel use to 1997/98. Power station fuel use is likely to develop broadly as shown in Table 6.3 which indicates that, while the total use of fossil fuels will remain broadly stable and the use of heavy fuel oil and Orimulsion (a derivative of Venezuelan heavy shale oil) will continue at a modest level, there will

Table 6.2 British Coal contract prices and tonnages to power stations (England and Wales)

	Average pit-head price £/GJ (1992/93 money)	*BC tonnage (million tonnes)*
1989/90	2.12	75
1990/91	1.89	70
1993/94	1.51	40
1994/95	1.46	30

Source: Author. Prices for 1989/90 and 1990/91 derived from Table 7.2 of White Paper (Cm.2235).

be a further major increase in the use of gas in CCGTs – with capacity building up to 14 GW by 1998, virtually all of which is already committed. This will put further downward pressure on coal use, which, by the end of the period, will be largely restricted to the contracted tonnages under the former BC contracts and with the smaller independent UK suppliers. The scope for imported coal will remain limited at this time. Most coal will be supplied by RJB (Mining) which secured 70 per cent of the coal industry, and most of the contracted business, when BC's mining assets were privatized by trade sale in 1994.

All projections of power-station fossil fuel use are subject to uncertainty from such factors as electricity demand, the performance of nuclear stations, and whether planned CCGTs come into operation on time. But the general shape of demand to 1997/98 is made more predictable by the 'take-or-pay' nature of the coal contracts, the large element of 'take-or-pay' in the CCGT gas supply contracts, and the 'must run' regime for nuclear stations. However, from 1998 the pattern of fuel use becomes more unpredictable. Firstly, with the planned ending of the RECs' franchises and the introduction of competition in the domestic market, it is unlikely to be possible for the UK coal industry (principally RJB Mining) to secure new 'take-or-pay' contracts for large tonnages incorporating premium prices. UK coal producers are likely to experience greater exposure to international coal price and currency movements. The major generators may also wish to diversify away from RJB, and/or to secure lower prices. Some increase in coal imports may occur if some UK coal is uncompetitive.

Table 6.3 Fuel used in power generation: major power producers 1994 and projection for 1997/98 (mtce)

	1994 (actual)	*1997/98 (projection)*
Coal	61	40
Oil[1]	6	5
Gas	15	40
Total fossil fuels	*82*	*85*
Non-fossil[2]	38	40
TOTAL	*120*	*125*

Source: Projection for 1997/98: author.
1 Including Orimulsion.
2 Nuclear, hydro, renewables and net imports over 'French Link'.

Secondly, the loading of CCGTs could fall if increasing competition in both electricity and gas markets (BG's monopoly in the domestic market is also planned to end in 1998) led to arbitrage between these markets, and the diversion of some gas supplies from power generation to other gas markets, or if gas prices rose significantly relative to coal. Thirdly, the regulation of SO_2 and CO_2 emissions could pose an increasing threat to the operation of the ageing stock of coal-fired stations. Specifically, the review by the pollution inspectorate (concluded in March 1996) of air pollution from power stations has greatly reduced the limits for SO_2 emissions in England and Wales. By 2001, these limits will be less than a third of the actual emissions in 1994/95. As additional FGD equipment is unlikely to be installed and new clean-coal technology is not yet commercially attractive, the new SO_2 emission limits will almost certainly further reduce the market for power station coal (particularly higher sulphur UK coal) progressively after 2000, and lead to additional gas-firing.

By 1998, the use of coal – virtually all of it UK coal – will almost certainly have been halved compared with 1989/90 (the last year before privatization) and the price of UK coal will be little more than half the 1989 price level in real terms – in late 1995 it is already a third lower. The use of natural gas in power stations, negligible in 1989/90, could well exceed the consumption of coal by 1998, and thereafter become the largest primary fuel source for power generation. In the (very) unlikely event of a recovery of coal's market share in the longer term, any increased supply would have to come from coal imports, in view of the massive reduction in UK deep-mined capacity which has taken place – and which is effectively irreversible.

Conclusions

It is important to consider to what extent these major changes in fossil fuel supply to UK power stations have been the result of ESI privatization. In order to come to a view on this let us consider some alternative scenarios for the 1990–95 period, assuming in each case the same economic conditions – in particular the same costs of UK coal and the same prices of imported coal and of natural gas.

If the former publicly-owned electricity undertakings in England and Wales had been privatized as a number of vertically integrated electricity supply organizations then, in the absence of large direct Government subsidies to UK coal (allowing BC prices to be aligned with imported coal), there would probably have been a substantial increase in coal imports and only limited use of gas. A similar outcome would probably have arisen if a unitary CEGB had remained in the public sector but had been allowed by the Government to exercise much greater freedom to choose its most economic sources of fuel. However, if it had been Government policy to provide sufficient subsidies to allow BC's prices to be aligned to those of imports, then, whether or not a unitary CEGB had been privatized, there would probably have been some increase in coal imports (with the aim of obtaining some diversity of coal supply), but a larger continuing use of UK coal and only limited use of gas.

There are always difficulties with such hypotheses, but they illustrate the way in which the radical changes in power-station fossil fuel use, which followed electricity privatization, have not been caused exclusively by privatization as such, but rather by the interaction between Government policy towards the coal industry – especially the political agenda to destroy the power of the NUM – the particular structure and regulatory framework of the privatized electricity industry and of British Gas which gave rise to the 'dash for gas', and conditions in the energy market generally.

As a result, five years after privatization, the ESI now has a greater diversity of fossil fuel supplies, although its coal supplies remain overwhelmingly in the hands of one supplier (RJB Mining) as a result of the way in which BC was privatized, at least to 1998, and the security of fuel supply cannot realistically be threatened by the NUM. Also, the 'dash for gas' has created significant headroom in meeting SO_2 emission limits and CO_2 containment targets for the 1990s and will thereby facilitate further environmental improvements in the longer term.

Two considerations, however, might lead to doubts as to whether the best outcome has in fact been achieved. Firstly, the 'dash for gas' was of a scale and speed which led to the excessive use of resources on new CCGT plant to a degree which was unjustified either in terms of total capacity requirement to meet demand or to minimize the total avoidable costs in the electricity system. Without the 'dash for gas', the phasing of the contraction of the UK coal industry might have been less draconian. Secondly, whatever the cause, the precipitate nature of the coal industry's contraction led to large-scale public expenditure (to support BC redundancy payments) and social costs which appear difficult to justify in cost/benefit terms. Arguably, at least some of the public money spent on redundancy and social security support for miners and their families might have been better spent in keeping more lower-cost pits in production (at least for a time) and enabling the rapid productivity improvement to reduce costs further.

Finally, there is an element of strategic doubt. The probability is that the already much-reduced UK coal industry will decline further over the next decade. As major investment to access new reserves is unlikely, currently accessible deep-mine reserves will be progressively exhausted, and environmental opposition will make it difficult to obtain replacement opencast sites. Moreovoer, the new SO_2 emission limits will severely restict the use of high-sulphur UK coals after 2000. In addition, by late-1995 it was clear that no new nuclear stations would be built (all plans to do so had been dropped) and that nuclear generation will decline as the older stations reach the end of their operating lives. The electricity industry therefore faces the long-term prospect of further increasing its dependence on natural gas, supplemented by imported coal to the limited extent to which coal use will be compatible with increasingly stringent environmental regulation. And the form of competition being developed may render obsolete the traditional structure of long-term contracts in favour of greater reliance on spot markets for fossil fuels. Whether these developments will lead to higher UK energy prices or reduced supply security will depend on the state of the international energy market at the time.

This is a further reminder that, beyond all the most direct and immediate determinants of policy, there is also the pervasive influence of the current conventional wisdom or view of the world. In many ways, it is difficult to imagine the changes in power-station fossil fuel use in the period 1990–95 taking place instead during the period of high oil prices of the 1970s when security of supply was often equated with indigenous supply and when gas was thought of as a high-value fuel, to be reserved for premium uses rather than bulk combustion. By contrast, following the collapse of oil prices in 1986, the period of the run-up to and following electricity privatization has been characterized by low and generally falling fossil fuel prices in real terms, and by expectations of a continuing abundance of fossil fuels, particularly gas. With strategic security of energy supply no longer a central concern of energy policy, the way was free for policies based on market forces and competition.

7

Nuclear Power Under Review

Gordon MacKerron

Introduction

The clearest failure in the government's privatization of the electricity supply industry was the abandonment in 1989 of the attempt to privatize nuclear power. Government had sponsored and nurtured nuclear technology since its early development in the 1950s, and all Governments in the UK have been strong supporters of the nuclear industry. When plans to privatize the ESI were developed after 1987, the Government also became committed to a competitive structure for the new private ESI. Early on in the privatization process, commentators began to suggest that nuclear power would not be capable of privatization in the competitive environment proposed.[1] Government however – apparently unaware of the unfavourable and highly risky economic status of British nuclear power – persistently tried to squeeze the nuclear industry into the new ESI structure.

Eventually, however, Government recognized the impossibility of fitting nuclear power into the new structure proposed for England and Wales,[2] and withdrew the Magnox stations from the sale in July 1989, and Sizewell B plus the AGRs in November 1989.[3] A moratorium on new nuclear construction was declared until a Nuclear Review was held, and this was promised for 1994. This chapter explains how this process happened, and examines the influence that the attempt to include nuclear power in the sale had on the structure of the privatized industry. It then turns to the complex questions of the financial protection of the still-public nuclear sector, especially via the fossil fuel levy, and the improved performance of the industry in the last five years. It then examines the Nuclear Review that government undertook in 1994/95, and the proposals that have emerged in the subsequent White Paper in May 1995,[4] especially for the renewed attempt – planned for mid-1996 – to privatize the AGRs and the Sizewell B PWR.

A Brief History

From the time of the Suez crisis of 1956 until the advent of the Conservative administration of 1979, Governments perceived nuclear generation as cheap and highly desirable. The CEGB, although initially hesitant, regarded nuclear power as its first choice for new capacity from the 1960s onwards. Early military developments had a very large influence on the civilian nuclear industry, and the two plutonium production plants built at Calder Hall and Chapelcross in the 1950s became the basis for the first civilian programme of nine Magnox stations, completed in the 1962–1971 period.[5] These stations had two reactors each and were built to different designs. They were physically large structures using natural uranium as fuel, and while they proved fairly reliable in operation, they carried with them very large 'back-end' costs – spent fuel management and decommissioning. Because these were mostly long-term costs, and because it was initially thought that the plutonium they would produce would have a high value as a fuel for fast-breeder reactors, these back-end costs did not appear to affect the economics of Magnox to any great degree up to the 1970s.

From the early 1960s a perception grew that reactors using enriched uranium (in which the proportion of naturally occurring uranium-235, 0.7 per cent, is raised to 2–3 per cent via 'enrichment') would offer better economy due to the possibility of more compact design. At the same time, enriched uranium became available in Europe. Britain developed a new design, using enriched uranium, and this became known as the Advanced Gas-cooled Reactor (AGR). This indigenous design came under threat from the emerging American water-cooled reactors, of which there were two main variants, the Boiling Water Reactor (BWR) and the Pressurized Water Reactor (PWR). In the apparent competition between the British AGR and American designs which was held in 1965, it was the BWR which offered the immediate threat to the indigenous design, though subsequently the PWR became more successful internationally. The competition staged between British and American designs was more apparent than real, and the Government decided to favour the AGR as the basis for the next nuclear programme.[6] This led to the hasty ordering of the Dungeness B reactors, which took 22 years to complete, and four subsequent AGR stations were ordered in the later 1960s. All suffered significant cost escalations and lengthy construction delays.

In the 1970s disillusion with the AGR design became general, and while the CEGB became keen in the early 1970s to build PWRs, the need for new plant was limited and it proved difficult to overcome technological nationalism in nuclear power. In 1978, however, a deal was struck with the Labour Government whereby the CEGB would be allowed to establish the option of building PWRs in the UK as long as two further AGRs were also ordered.[7] Thus the Torness and Heysham 2 AGRs started construction in 1980, further development work having been done on the AGR design. Preparations were also made to design a Westinghouse-based PWR for construction at Sizewell.

When the Conservatives took office in 1979, they were inclined to encourage nuclear power much as other Governments had done, with the added support of Mrs. Thatcher's enthusiasm for high-technology activities. But the Conservative Government had another powerful political motive for supporting new nuclear construction. Competition for new ESI investments was, by the late 1970s, exclusively between coal and nuclear, and every new nuclear station would, in a situation of low electricity-demand growth, displace coal from its main market. The Conservatives blamed the miners' strike of 1974 for the loss of the General Election of the same year, and were keen to cut down the economic and political power of the miners (see Chapter 6). They therefore quickly adopted a policy of major nuclear construction, announcing in late 1979 that there would be a programme of ten large PWRs to start construction annually from 1982.[8] A major, generic Public Inquiry would be held into the proposal for the first PWR at Sizewell: thereafter, it was assumed, further PWRs might be built without difficulty.

This ambitious timetable immediately ran into difficulties. The first UK PWR design was very expensive, highly complex and too distant from its US template (the so-called SNUPPS plant design) and a virtually complete new start had to be made in the design process. This delayed the opening of the Public Inquiry by a year, and it finally got under way in

1982. The Inquiry admitted evidence on a very wide range of subjects (including nuclear weapons and energy policy in general) and it proceeded at a leisurely pace. It did not complete its hearings until 1985, and the Inspector's report was not published until 1987,[9] delayed – it seemed – mainly by adverse public reaction to nuclear power in the wake of the Chernobyl accident of 1986.

The report was highly favourable to the proposed Sizewell B. It argued that there was only a 1 in 40 chance of a coal-fired project proving cheaper than the Sizewell B PWR, and that it would probably save overall system costs if new PWRs replaced old fossil stations even without any 'need', on demand-growth grounds, for new plant (see Chapter 2). By 1987 much of the data in the Report was out of date and most of the developments in the mid-1980s had been adverse to nuclear economics. Not very surprisingly therefore, the economics of Sizewell B have proved much poorer than the Inspector expected. Its capital costs escalated (in real, inflation-adjusted, terms) by a minimum of 35 per cent,[10] and it will generate (at best) at costs roughly double the spot price for electricity established in the Pool.[11] Approval for this project finally came in April 1987, almost eight years after the Conservatives' new PWR policy was announced. A proposal to build a second PWR at Hinkley Point also went to a Public Inquiry starting in 1988, but before the hearings were over, the nuclear stations had been withdrawn from the ESI privatization and the Secretary of State for Energy had announced a moratorium on new nuclear construction. The Hinkley project was given planning permission but no construction was sanctioned pending the Nuclear Review. Thus plans for rapid growth in nuclear power during the 1980s had come almost to nothing by the time of electricity privatization: only a single PWR was being built and it would not be fully commissioned until September 1995.

The Economics of Nuclear Power in Britain

Until the problems of nuclear privatization emerged in 1989, it had been widely assumed that nuclear power was a highly economic source of power generation for Britain. However, this perception depended heavily on a particular, and flawed, presentation of cost comparisons. For an overall assessment of the economics of nuclear power against alternative options, it is necessary to conduct a full cost analysis, in which all the capital charges are added to operating costs at the prices of one year. The other useful form of analysis would be to look at the avoidable costs (those which would cease in the event of a close-down) and check whether or not nuclear costs were below the selling price for electricity, and therefore whether nuclear plants were worth keeping open. From 1979 onwards the CEGB started to publish figures designed to show the cheapness of operating nuclear stations, but it conducted neither a full-cost nor an avoidable-cost exercise. Instead it concocted a hybrid analysis, which appeared to constitute a full-cost analysis but in reality was nothing of the sort.[12] In comparing Magnox and AGR stations against coal-firing, the CEGB added fuel cost in current money prices to capital charges at historic values. Because the economics of nuclear power were dominated by capital, and coal-firing costs were dominated by fuel, the result was a systematic bias in favour of nuclear. These calculations were not (and could not be) the basis for any decisions, but they did have the advantage from the CEGB's point of view of creating a climate – when new investment in the Sizewell PWR was about to be discussed at a big Public Inquiry – in which nuclear economics seemed on a sound footing.

Re-calculations of the CEGB's figures on a full-cost basis showed that even though fossil fuel prices were at unprecedentedly high levels in the 1979–81 period, coal-firing was generally cheaper in whole-lifetime terms.[13] In other words the record showed that both Magnox and AGR nuclear investments had been, on a full-cost basis, an economically unsound decision. CEGB figures produced for 1982/83 and 1983/84 suggested that *to date* nuclear had not been cheaper than coal (by a margin of some 7 per cent to 9 per cent) but that assuming high coal prices in future, the full costs over the whole lifetimes of plants would show nuclear to have been the cheaper option.[14]

Developments later in the 1980s completely undermined this hope: the economics of nuclear power became significantly worse than the economics of coal-firing. Perhaps as a consequence of this, no further relative generating-cost figures were published before privatization. On the coal side, fuel prices stabilized and then in real terms began to fall (even though UK coal prices remained well above internationally traded levels). For the nuclear stations, costs began to escalate significantly. First, the completion of three of the CEGB's first-generation AGRs continued to be delayed and led to further escalations in capital costs for those stations. When operation was achieved, three of the AGRs appeared to be chronically unreliable, and the costs of AGR generation – with high capital costs spread over relatively few units of output – were clearly very high. Second, the costs of the 'back end' escalated sharply, and this applied most powerfully to the Magnox stations. This escalation was partly for technical and partly for commercial/political reasons. BNFL had an effective monopoly in the provision of spent-fuel management services (essentially fuel reprocessing and waste management) and had always provided these services on a cost-plus, pay-as-you-go basis. The dominant back-end cost was reprocessing, and for Magnox stations the real unit cost of reprocessing doubled between 1979 and 1985,[15] partly because of a need for a very costly refurbishment of the ageing spent-fuel reprocessing plant at Sellafield. As decommissioning costs became defined for the first time in the early 1980s it was estimated that the undiscounted cost of full decommissioning for all CEGB and SSEB stations would amount to over £3 bn.[16]

Because there was no public information on nuclear costs after the 1983/84 financial year, it was impossible to gauge by how much the economic status of operating nuclear stations had worsened by the time of the run-up to privatization. However, it should have been clear that the operating costs and financial risks of nuclear power were substantially higher than fossil-based alternatives. Even if avoidable costs justified continuing operation, unavoidable costs were high, and someone would have to pay for them.

The First Attempt to Privatize Nuclear Power

Thus when the privatization issue surfaced, the total costs of running nuclear power stations were substantially higher than for coal-fired generation. Further, many of the costs involved were very difficult to estimate accurately and involved large risks which were difficult to place with any specific party – this was especially so with the very large future liabilities for reprocessing, waste management and decommissioning, which applied to all British reactors, but above all to the Magnox stations. In this context, it was going to be virtually impossible to privatize nuclear power in a structure which would no longer allow any automatic cost pass-through. This virtual impossibility was argued by outside commentators as early as 1987.[17]

Government did, however, recognize that nuclear stations carried large financial and technical risks – especially for back-end liabilities due to be discharged in the future – and this recognition was the largest single influence on the proposed structure of the new generating companies.

While a system of some 60,000 MW (England and Wales) could have been broken into five or six generating companies – a number which would have led to good possibilities of competitive behaviour between generators – the Government decided that the financial hazards of nuclear power would need a large shelter. Thus in England and Wales, a generating company with some 70 per cent of total capacity was thought necessary to shelter the 16 per cent or so of capacity that was nuclear. This was the company that eventually became National Power. However, having established a giant, the Government was unwilling to allow it the huge dominance that creation of two or more other generators would have established. The remaining 30 per cent of generating assets in England and Wales were therefore to be owned by a single other company – eventually PowerGen. In Scotland, as elsewhere in the privatization, things were to be different: it was originally proposed that the larger of the two integrated Scottish companies – Scottish Power – would simply retain the nuclear stations that its predecessor, the SSEB, had previously operated.

While the Government did not understand quite how expensive the costs of existing nuclear capacity had become, it did recognize in 1987 and early 1988 that it might be difficult to persuade the privatized RECs to contract for its desired *new* nuclear capacity unless the RECs were compelled to do so. In addition to the financial shelter of a large generator, the Government therefore also proposed a non-fossil fuel obligation (NFFO) which would specify quantities of nuclear (and later renewable) electricity that the RECs would have to purchase. It was originally expected that this obligation would be set at levels which would allow the completion of a proposed 'small family' of four PWRs. The whole of the CEGB's Public Inquiry case for the proposed Hinkley Point C PWR in 1988 was based on this premise.[18]

As the high operating costs and risks of current nuclear operation became more apparent to Government, a fossil fuel levy (FFL) was introduced in England and Wales in late 1988 as a way of financing these expected excess costs of non-fossil-based electricity. It was to be a flat-rate tax on all retail sales of fossil-fuelled electricity, the proceeds of which would compensate the RECs for the higher costs of nuclear and renewable electricity purchase. The haste with which the levy was invented helps account for its particularly opaque and rigid character: payments from it to Nuclear Electric were fixed in money terms for each of eight financial years from 1990/91, and were kept secret until late 1992. It then emerged that Nuclear Electric was due to receive over £9 bn of consumers' money over the eight years, irrespective of its cash needs.[19]

Finally, in the structure of special help for nuclear power, a provision (which became Schedule 12 of the Electricity Act 1989) was to make available up to £2.5 bn to the owners of nuclear plant for unanticipated escalation in back-end expenses in the future. The attempt to privatize nuclear power therefore led to plans for a series of distortions in the market.

The main negotiation in this first attempt at nuclear privatization was over the terms on which National Power would take on the English and Welsh nuclear stations. This in turn depended on negotiations between the shadow National Power (still technically part of the CEGB) and BNFL for fixed-price contracts on fuel-cycle services. The negotiations between National Power and Government proved difficult and, in the end, fruitless. In this aborted run-up to nuclear privatization there were two simultaneous processes in the estimation of the costs of nuclear power in the private sector. One was, for the very first time, a hard-headed look at the real costs and risks associated with nuclear power operation. The second was a bargaining game, in which National Power had an interest in inflating the apparent costs of nuclear power in order to secure for itself the best possible commercial terms for taking on the nuclear stations. It is difficult precisely to distinguish between the 'real' and the 'bargaining' cost increases, but there is no doubt at all that the 'real' increases were very substantial.[20]

Among the main increases in the components of nuclear cost 'discovered' in the 1987–89 period, the following are the most important:

- the anticipated cost of decommissioning an average Magnox station approximately doubled from £312 m to £600 m, mainly because of uncertainties attaching to a process that had yet to be commercially established;[21]
- once BNFL calculated the cost of their own decommissioning to the point of returning their sites to greenfield status, the total undiscounted bill escalated approximately eleven-fold, from £438 m to £4605 m[22] (although a remarkable overnight increase, this was not a major problem for BNFL, which had contractually arranged that its customers would pay for almost the whole of BNFL's decommissioning bill);
- the new fixed prices for BNFL fuel cycle services were to be some 20 per cent higher to reflect risks[23] (bearing in mind that the most expensive item, Magnox reprocessing, had already doubled in unit

price in the early 1980s); and

- in the appraisal of the costs of new PWRs, the cost of capital was doubled from 5 per cent to 10 per cent, and a whole range of other 'private-sector-prudent' assumptions were applied to the calculation of the prices that would need to be paid for PWR output.[24]

Lord Marshall (who resigned from Chairmanship of the CEGB when nuclear privatization failed) tried to show how PWR costs could almost triple if 'private sector' input assumptions replaced public sector numbers.[25]

The fact that it was now in National Power's interest to talk up nuclear costs (when for decades in the public sector the same people had an institutional incentive to make these costs seem as small as possible) means that not all the above increases can be taken at face value. Nevertheless the Government clearly believed that the bulk of them were real enough, and it responded by withdrawing the nuclear assets from the privatization in two stages.

First it became evident that the various escalations in Magnox back-end costs were very large. Up to early May 1989, nuclear liabilities were expected to be between £3.5 bn and £4.5 bn,[26] a range which Government thought manageable for National Power. In May, however, National Power made it clear that the various cost escalations at the back end meant that realistically the liabilities for nuclear stations would be closer to £10 bn, surrounded by a wide range of uncertainty. Virtually all of this huge escalation was accounted for by the Magnox stations, and, as this would constitute an unsupportable burden for National Power's opening balance sheet, the Magnox stations were withdrawn from the sale on 24 July 1989.

At this point it was still believed in Government that the AGRs (plus Sizewell B and any future PWRs) could be sold as part of National Power. The first difficulty in the negotiations between Government and National Power concerned prospective selling prices for PWR electricity. The Department of Energy had worked on a private-sector price for PWR electricity of around 5p to 5.5p per kWh, pending advice from National Power (which did not come until October 1989).[27] Such a price level was high in relation to the 2p to 3p per kWh range of the Hinkley Point C Public Inquiry but Government believed it was still manageable in the context of privatization. However, the prices quoted by National Power came in at a range of 6.37p to 8.27p per kWh, even excluding interest during construction. Further negotiations made little progress, least of all with risk-sharing:

risks were to be borne overwhelmingly by consumers and/or Government. AGR prices – surprisingly neglected by Government, fixated as it still was on a programme of future PWRs – were also to be at least 5p to 6p per kWh. In these circumstances, where high and barely quantifiable risks were being reflected in very high fixed prices, there was clearly no real alternative to withdrawing Sizewell B plus the AGRs from the privatization. This was done on 9 November 1989, together with the announcement of a moratorium on nuclear construction and a Nuclear Review to be held in 1994.

By this time the Government was committed to a tight privatization timetable, and the process of re-organizing the larger part of the former CEGB into two generating companies was advanced. To have tried to split either or both of the shadow generators into yet more companies would have seriously delayed the timetable, as well as being organizationally complicated. The Government, with a new 'fixer' in charge of Energy (John Wakeham), therefore simply endorsed the generating structure as originally proposed, with the simple deletion of nuclear power from National Power's portfolio. In Scotland, where lower nuclear costs and the integrated utility structure might have allowed privatization of the Scottish nuclear stations to go ahead, it was instead decided to follow the English example and withdraw the three nuclear stations in Scotland (two operating AGRs and a newly shut Magnox) from the sale. In some haste, therefore, two new public sector nuclear companies were set up, Nuclear Electric for England and Wales, and Scottish Nuclear for Scotland. For England and Wales the NFFO was to remain but all that it would now do, in the nuclear area, was to ensure that all nuclear electricity from Nuclear Electric was purchased by RECs, rather than to compel the construction of new PWRs.

The nuclear debacle therefore had a critical effect on the structure of the new privatized industry. Restricted competition in generation appeared to be the price necessary to pay for successful nuclear privatization: in the end, the duopoly in generation was left without any rationale at all. The Secretary of State for Energy responsible for the final stages of the privatization later claimed that, but for the need to try and accommodate nuclear power within the privatization, a generating structure of around five companies would have been instituted.[28] Many of the problems of regulation in the new ESI have precisely stemmed from the dominance in the overall market of National Power and PowerGen.

The new structure for nuclear power was also unpromising. Two public-sector, nuclear-only utilities were created with poor financial prospects and a restriction that they could invest only in nuclear power stations (though not until 1994 at the earliest). Even the completion of the part-built Sizewell B PWR was publicly scrutinized, and although it was decided to continue with the project, the official cost calculations suggested that the economics of continuing with Sizewell was (despite sunk costs of nearly £1 bn already) only a little better than marginal.[29] The provisions for decommissioning and other long-term liabilities built up by the CEGB since 1976 were largely lost: they had mostly been invested in non-nuclear assets that had gone into the private sector with National Power, PowerGen and the National Grid Company. The fortunes of nuclear power in Britain were, in November 1989, by some distance at their lowest ebb.

SUPPORTING PUBLIC SECTOR NUCLEAR POWER: THE FOSSIL FUEL LEVY AND THE NFFO

Despite the major reversal in the fortunes of nuclear power experienced in 1989, it had not disappeared, and indeed it had too large a share in generation (20 per cent, taking Britain as a whole) to be disposed of in the short term. For the new public sector companies, Nuclear Electric and Scottish Nuclear, ways still had to be found to keep existing nuclear assets working. The first question was whether or not this support should be purely short-term – after all, if the generating costs of nuclear power were so much higher than of the fossil alternatives, surely the logical thing to do is to close nuclear power down as soon as there is adequate alternative capacity?

The gap between the cost of nuclear and fossil-generated electricity was substantial in 1990/91. In that year, average operating costs for Magnox and AGR stations were 4.5p/kWh,[30] against contract prices for base-load electricity of around 3p/kWh. The main cost escalations had centred on Magnoxes, and AGRs had prospects of generating more cheaply if their availabilities could be improved. The main question therefore focused on the Magnoxes, which were old and mostly small. The argument used in defence of the continued operation of the Magnox stations was that while their operating costs are high, their avoidable costs are low. In other words, while for most generating units operating costs cease when the stations close down, for Magnox the bulk of these costs cannot be avoided. Thus Nuclear Electric argued in 1992 that the avoidable costs of Magnox were as low as 1.2p/kWh,[31] well below the Pool or contract price for electricity in the market. This means that almost 75 per cent of the operating costs of Magnox are essentially fixed – they will be incurred whether or not the stations are running. This is understandable in the case of decommissioning – clearly a fixed cost – but decommissioning constituted only a tiny proportion of unavoidable costs. The real problem lay at the back end of the nuclear fuel cycle, where reprocessing and waste management amounted to over 2p/kWh, and where hardly any of this cost would apparently be avoided by shutdown.

This inflexibility in the back end of Magnox operations was at one level purely financial – Nuclear Electric was negotiating contracts with BNFL which allowed very little reduction in cost if the throughput of spent fuel fell. But from a national resource perspective this was unimportant: what mattered was not the contracts but whether or not real resources could be saved by closing the Magnoxes down. The consultants who examined this issue for Government at the time of the pit-closure dispute in 1992 came to the conclusion (though some of their figures were withheld because of 'commercial sensitivity') that *to Nuclear Electric* the extent of unavoidable Magnox costs was indeed as high as the company claimed.[32] What the consultants did not do was to examine the internal cost structures of BNFL to determine whether the fixity of reprocessing operations is a technical fact or a reflection of BNFL's monopoly powers. Such an investigation has never taken place in the public domain and the underlying technical and economic magnitudes cannot therefore be checked. In the meantime Nuclear Electric has continued to argue that Magnox avoidable costs remain on average around 1.4p/kWh,[33] a figure which clearly justifies continued operation if validated.

Government, therefore, in the face of these arguments that appeared to justify long-term operation of nuclear assets despite their high costs, needed to find long-term financial support measures to compensate for the fact that accounting costs of nuclear power would clearly be a lot higher than the price at which a market would be willing to buy nuclear electricity. Different approaches were taken as between England and Wales, and Scotland. In England and Wales the devices of the NFFO and FFL – originally invented for private nuclear ownership – could still be used; for Scotland a different approach was taken.

The nuclear situation in Scotland was quite different from that in England and Wales. First, nuclear electricity was much more important in Scotland, accounting for some 40 per cent of generation (and rising), against less than 17 per cent in England and Wales. Second, while the CEGB had owned eight Magnox stations and five AGRs, the old SSEB possessed only one Magnox plus two AGRs. The Magnox – the small and relatively efficient Hunterston A – had been closed on the eve of privatization, leaving Scottish Nuclear with two operating AGRs. These were two of the more successful AGRs, Hunterston B from the first generation, and Torness from the second round of ordering in 1980. Relatively unencumbered by the large liabilities affecting Magnox, and owning two AGRs with relatively good operating performance, the long-term financial health of Scottish Nuclear was potentially less chronic than that of Nuclear Electric.

Scottish Nuclear did not therefore need the explicit injection of current funds that were deemed necessary for Nuclear Electric through the fossil fuel levy. Instead, £1.37 bn of the outstanding £1.6 bn debt (mostly relating to the building of the recently completed Torness AGR) was written off; £716 m of the £2.5 bn of Schedule 12 moneys was earmarked for Scotland; and contracts were signed for the sale of all Scottish Nuclear output in a Nuclear Energy Agreement stretching from 1990 to 2005. This allocated 74.9 per cent of Scottish Nuclear (SN) output to Scottish Power and 25.1 per cent to (Scottish) Hydro-Electric. The price payable for this electricity was fixed at a premium level (averaging about 3.3p/kWh in 1990 money) for the four years to 1994, sliding downwards for the four years to 1998, at which point the price would be based on the average price in England and Wales for base-load electricity.[34]

These Scottish arrangements have attracted less attention than those in England and Wales, mainly because the subsidy (most of it contained in the premium price) has been less visible. However, they have introduced considerable rigidity into the Scottish market, as the contracts specify that all Scottish Nuclear output must be bought by the Scottish utilities. This means that as the output of the AGRs in Scotland has risen from just over 12 TWh in 1990/91 to 16.8 TWh in 1994/95[35] (now supplying some 43 per cent of Scotland's electricity) so the domestically sold output from other Scottish stations necessarily falls, irrespective of relative costs. However, this has limited impact on consumers, as Scottish Nuclear recovers its premium on a fixed number of units sold; marginal nuclear output is sold at CCGT-related prices.[36]

In England and Wales the mainstay of the financial support to Nuclear Electric has been the proceeds of the fossil fuel levy (combined with the NFFO requiring RECs to buy nuclear output). While the NFFO applies to all non-fossil forms of generation, it was invented as a device to get round the problem of the excess costs of nuclear power, with renewables added in as an afterthought (renewable NFFO issues are fully covered in Chapter 8). This section therefore concentrates on only the nuclear-specific aspects of both the NFFO, and of the FFL which is designed to fund it.

The NFFO is a simple device which allows the Secretary of State (originally for Energy, now for Trade and Industry) to require RECs to buy specified quantities of non-fossil generated power. In the nuclear case, the rule is that RECs must buy all output offered to them by Nuclear Electric (NE). The compensation for the RECs is that they pay only a Pool-related price for their nuclear electricity to NE, who are then paid the great bulk of the proceeds of the fossil fuel levy to compensate them for the higher costs that nuclear power incurs compared to competitive forms of generation.

The fossil fuel levy is a tax on all retail sales of electricity (currently 10 per cent), which must be collected on all electricity sales (with the sole exemption of self-generation if more than 50 per cent of self-generated electricity is consumed on the same premises as the generation). Most of the collecting duty falls on electricity suppliers who then pass the proceeds on to the Non-Fossil Purchasing Agency, which in turn pays NE. It could last only until 1998 – the European Commission classed it as a state aid and required that it cease after eight years. The magnitude and precise purpose of the FFL was obscure in the first two years of privatization.[37] Although the sums to be collected had been specified in advance, they had not been made public, and the purpose of the levy was widely – but almost wholly inaccurately – thought to be to allow NE to pay for the decommissioning of the nuclear stations which it had inherited from the CEGB. Even the President of the Board of Trade repeated the error in October 1992 when he announced in the Commons that the levy was to pay for 'the decommissioning of old and unsafe stations'.[38]

This view contained two misconceptions. The first was that the main expenses arising from the 'old and unsafe' stations were for decommissioning. In fact only three nuclear stations have been closed and the intention is to postpone the bulk of expenditures on decommissioning for a century and more. Only 1 per cent of levy proceeds have to date been spent on decommissioning (as Table 7.2 below shows). Significant amounts of money have been spent on discharging liabilities arising from older stations (mainly Magnox) but these have been almost exclusively in the areas of fuel reprocessing and waste management. The second misconception was that the levy was somehow reserved for inherited liabilities. In fact there was no hypothecation, and the levy simply constitutes a large supplement to NE's trading income. Because the rate at which liabilities can be paid off is limited by the capacity of BNFL to reprocess spent fuel, a very large part of the levy to date has necessarily been spent on current business activities, especially paying for NE's investment programme. The Finance Director of NE gave the clearest explanation of the true purpose of the levy: it was to pay for inherited liabilities, *and* to ensure that the company remained cash-positive.[39] This meant that electricity consumers, rather than the Treasury and taxpayers, were made responsible for ensuring that NE did not lose money.

Only when the House of Commons Trade and Industry Committee investigated the pit closures proposed in late 1992 did any public information become available on the payments due each year to NE under the FFL (Table 7.1). NE could not obtain these very large payments (£9.1 bn over the full eight years) automatically – the levy would only be fully payable if a minimum level of electricity were generated each year (also shown in Table 7.1). These levels of output were intended as a demanding target, though in the event, NE's success in operating AGRs means that the qualifying levels of output have had no relevance. The levy is payable only on the 'levy-related outputs' shown in Table 7.1, and electricity generated above this level must be sold by NE for whatever price it can get in the market.

Table 7.1 Planned annual levy payments to Nuclear Electric

Year	*Levy (£m 1992)*	*Levy-related output (TWh)*	*Actual output (TWh)*
1990/91	1290	38	45.0
1991/92	1306	38	48.4
1992/93	1282	39	55.0
1993/94	1183	40	61.0
1994/95	1179	40	59.2
1995/96	1059	40	–
1996/97	967	42	–
1997/98	857	39	–

Sources: House of Commons Trade and Industry Committee *British Energy Policy and the Market for Coal* HC 237, 26 January 1993, Table 18, para 120; OFFER *Submission to the Nuclear Review* October 1994, Table 3, p 17; and Nuclear Electric *Report and Accounts 1994/95*, p 2.

The mechanics of the levy are that the DGES must, each year, fix the rate on qualifying (fossil-fired) electricity sales that will be needed to raise the pre-determined cash sums shown in Table 7.1. This is why the precise annual rate of the levy cannot be known in advance. In practice the levy has been in the range of 10 per cent to 11 per cent, and has been steady at 10 per cent for the years 1994/95 and 1995/96.

In the five years since April 1990, some £6 bn of consumers' money has been paid to Nuclear Electric from the levy. In 1990/91 levy income was over 50 per cent of NE's revenue: even by 1994/95, when NE's finances were a good deal healthier, 43 per cent of total income was derived from the FFL.[40]

To what uses have NE put this very large sum of money? One kind of answer is that as the levy is simply added in to NE's trading income it is impossible to ascribe specific expenditures to particular revenue streams. This is, however, a disingenuous answer and a simple analysis assists in the understanding of the uses of the levy. This is presented in Table 7.2, which shows the sums of money received each year from the levy by NE[41] and the amounts actually spent on paying for the discharge of inherited liabilities. These payments have been overwhelmingly for reprocessing spent fuel and subsequent waste management. The expenditures for decommissioning, however, are shown separately because of the misconception that the levy is only to do with decommissioning. As Table 7.2 shows, decommissioning expenditure has so far amounted to almost exactly 1 per cent of NE's total levy receipts. For the first four years shown in Table 7.2 the payments for historic liabilities were heavily concentrated on the Magnox stations.

It is clear from Table 7.2 that the results of the levy have been to make NE 'cash-positive' by almost £2.9 bn more than if the levy had only funded the discharge of historic liabilities. This very large sum has been put to two main uses. The first of these is allowing the company to complete Sizewell B together with other minor investments (£1.6 bn)[42] without recourse to any borrowing. This has obviously been financially very useful to the company.

The second use of 'surplus' levy money has been to help Government reduce its need to borrow. As much as £1.5 bn of levy receipts were, at 31 March 1995, held on deposit by Nuclear Electric with Government.[43] As this counts as public-sector money on the same basis as any other public funds, Government's need to borrow is correspondingly reduced. With Nuclear Electric's investment programme essentially completed in 1994, a

Table 7.2 Nuclear Electric: levy receipts, and payments for historic liabilities (£m, current money)

	1990/1	*1991/2*	*1992/3*	*1993/4*	*1994/5*	*Total*
1 Levy receipts	1195	1265	1280	1230	1251	6221
2 Payments made for historic liabilities	485	522	577	605	1097	3286
of which decommissioning	9	9	6	17	21	62
3 Cash remaining for current business (row 1–row 2)	710	743	703	625	154	2935

Sources: Nuclear Electric *Annual Reports and Accounts*

very high proportion of levy moneys now flow from consumers through Nuclear Electric's books and straight into Government central finances. The levy therefore now constitutes, in large measure, a tax on electricity consumers, rather than a means to keep a public company afloat or to help it pay for inherited liabilities.

This rather detailed consideration of the operation of the levy has shown its extraordinary consequences. Not only has it funded over £3 bn of payment for back-end liabilities, it has subsidized NE's current cash flow by some £2.9 bn (thus allowing it to finance Sizewell B without needing to borrow a penny) and provided an increasing source of useful cash for the Treasury to offset against general public sector borrowing needs. It is perhaps not surprising that the FFL has had few defenders and many harsh critics, including industrial users of electricity and – in a more muted way – OFFER.[44] Even the proposals to privatize part of the nuclear industry again in mid-1996 have not led to a complete demise of the levy.

THE PERFORMANCE OF NUCLEAR POWER SINCE 1990

The elaborate financial support structures for nuclear power clearly gave NE and SN some time in which to repair the performance of nuclear power, but repair was vitally necessary. At vesting of the two new public sector companies, Nuclear Electric stations were generating at

5.3p/kWh,[45] while the AGRs were achieving less than 50 per cent availabilities, and had operating costs of nearly 3.3p/kWh even in Scotland,[46] where performance had generally been better than in England and Wales.

The technical and financial performance of both public sector companies since 1990 has been remarkable. This has suggested that incentive structures can be more important than ownership in determining performance: the improvement in financial performance is comparable to that of the private sector electricity companies that were created at roughly the same time. These new incentives and pressures facing the nuclear companies have been the scrutiny involved in the 1994 Nuclear Review, and the greater transparency in nuclear activities being separately accounted for and compared directly with the new private sector generators, though the explanations for better operating performance are unlikely to reside solely in incentive structures (as argued below).

The most important single improvement made by both companies has been in the operating performance of the AGRs. Table 7.3 shows the operating performance of each reactor over the period 1989 to 1995. The AGRs belonging to NE have improved their performance from an average load factor of 42 per cent in 1989 and 1990 to 66 per cent in 1995 (with even better years in 1993 and 1994). Scottish Nuclear AGRs have also improved substantially, and Scottish reactors averaged 80 per cent load factor in 1995. These improvements are by a long distance the most important reason for the large improvement in the financial results of the two companies, because incremental output from AGRs incurs low incremental costs (almost entirely those in the fuel cycle, estimated by SN at 0.9p/kWh) while the companies obtain a selling price at least as high as the Pool price (roughly 2.5p/kWh). Given that both companies are effectively guaranteed markets for all the electricity they can produce, the contribution to profitability of enhanced AGR output is very substantial. Using the 1.6p/kWh profit suggested by the above figures means that an extra 20 TWh – roughly the increase in NE's AGR output – would improve pre-tax profits by over £300 m a year.

There are two obvious questions to be asked about the very large improvement in AGR performance: how did it happen, and can it be sustained? NE and SN tend to argue that the principal cause of better performance of the AGRs was the new incentive structures set up in both companies to reward improved financial performance. It is possible that this has made some difference, but it seems even more likely that much of the improvement would have happened in the early 1990s what-

Table 7.3 Annual load factors[1] of AGRs, 1989–1995 (%)

	1989	*1990*	*1991*	*1992*	*1993*	*1994*	*1995*[2]
Nuclear Electric							
Dungeness B	9	16	42	35	57	49	19
Hartlepool	39	47	47	67	82	75	74
Heysham 1	53	46	61	59	75	78	52
Heysham 2	58	22	38	73	86	85	93
Hinkley Point B	57	72	66	71	83	86	90
Average England and Wales	*43*	*41*	*51*	*61*	*77*	*75*	*66*
Scottish Nuclear							
Hunterston B	73	76	64	59	73	80	80
Torness	57	36	48	73	70	71	79
Average Scotland	*65*	*56*	*56*	*66*	*72*	*76*	*80*

Source: *Nucleonics Week* various issues

1 Load factor is the total electricity produced as a proportion of the total electricity that would have been produced if the station had operated at full design output all year round.

2 For 1995, Nuclear Electric figures are only for the first nine months, as the company argues that revealing data about operations would be prejudicial under the Financial Services Act. Scottish Nuclear does not appear to suffer from the same limitation.

ever structures had been in place. Management and engineering attention necessarily switched away from the abandoned programme of PWRs after 1989. Moreover, five of the seven AGRs had then only been in service for one to three years, the time when 'troubleshooting' activity would in any case normally occur. In the case of Nuclear Electric, some of the improved performance has been obtained from raising the power ratings of their stations towards design levels – in the five years to 1995, the power ratings of English AGRs have risen by over 900 MW. The rest of Nuclear Electric's improvements have been concentrated on succeeding in making some of their stations (especially Heysham 2 and Hinkley Point B) consistent producers of high output. However, doubts persist on the prospects for the other three stations in England – especially Dungeness B – to produce high output levels reliably. In Scotland the improvements in AGR performance seem to have been more consistent, and have been achieved entirely by raising output at power ratings that were close to design levels in 1989/90.

Whether or not the improvements in AGR output can be sustained is difficult to say. Four of the seven AGRs now have the ability to refuel while operating ('on-load refuelling'),[47] which should help sustain performance levels, though it seems doubtful that on-load refuelling can be achieved at the other three, Dungeness B, Hartlepool, and Heysham 1, which have always been the most problematic of the whole group. English AGRs also continue to show high levels of unplanned shutdowns by international standards[48] and this would continue to constrain operating performance. However, it seems likely that most of the AGRs will be able to perform in the 70 to 80 per cent range for some years to come.

Better AGR performance has mostly affected the income side of the equation – NE's income from electricity sales has risen from £1.01 bn in 1990/91 to £1.64 bn in 1994/95[49] – but both companies have been trying to cut costs. As in their private-sector counterparts, this has principally involved reducing their labour force. NE had 13,924 employees at Vesting, and at March 1995 had reduced this level to 8990, with further cuts planned. Scottish Nuclear initially increased its employment levels but has been cutting back in the last two years. In April 1990 employment stood at 1976, rising to 2100 in 1993 and falling to 1737 at March 1995. The impact on total costs is more limited than these figures imply however because both companies have introduced 'contractorization' to a marked degree – many of the functions previously performed by company employees are now put out to competitive tender, so that the company head-count exaggerates the extent to which there has been a reduction in the real quantity of labour applied to the same corporate work (See Chapter 9).

The other major cost-cutting attempt has been in the long-drawn-out process in which both companies have been negotiating with BNFL for large, long-term fuel cycle contracts. Negotiations opened soon after 1990, but final signatures were not obtained until March 1995. These are huge contracts – NE's is for £14 bn and SN's for £4 bn – covering 15 years' worth of fuel-cycle and waste-management services.[50] NE claims that the final terms with BNFL save the company some £1 bn compared to earlier contractual arrangements, mostly concentrated on the (more expensive) Magnox side,[51] while SN also claims large savings. In SN's case this saving has been concentrated on the terms offered for an expansion of AGR fuel reprocessing (the company had for some time been planning to build a dry fuel store as an alternative to reprocessing). Originally SN claimed that dry stores would have saved them £45 m annually.[52] The SN/BNFL contract is said to match the savings which SN would have had through use of dry-store technology. At least as important as these cost savings has been the distribution of economic risks between the generators and BNFL. Both NE and SN have stressed that as well as saving money, the new contracts also place significant increases in commercial risk on to BNFL in the event of unexpected cost escalation for regulatory or other reasons. These arguments are almost certainly correct, and reflect one of the necessary preparations for the second round of privatization, in which minimizing the financial risks for the proposed new private companies is a necessary ingredient of a successful sale.

The overall financial results of these various changes have been an impressive improvement in performance in both companies. NE's trading income has risen by over £600 m in the five years to 1995, while its operating costs have fallen by £100 m.[53] Scottish Nuclear's income has risen from £423 m to £580 m while its operating costs have fallen from £433 m to £398 m. NE made a loss of £423 m in 1990/91 even after receiving over £1.1 bn in levy payments; by 1994/95 it returned a profit of £1218 m, only marginally below its levy receipts for the same year (£1251 m). Scottish Nuclear lost £33 m in 1990/91 and had turned this into a profit of £150 m by 1994/95, and claims that it would have made a £31 m profit if it had received no subsidy via their price premium. These are real financial achievements and help explain why, in 1995, the Government felt able to announce a renewed attempt to privatize the relatively more profitable parts of the industry.

THE NUCLEAR REVIEW AND PRIVATIZATION

In the panic of November 1989 the Secretary of State for Energy announced – besides the withdrawal of all nuclear power from the privatization – a moratorium on new nuclear construction until the results of a major review of nuclear power had been held, and promised this for 1994. The Trade and Industry Select Committee urged Government early in 1993 to bring the review forward.[54] Government agreed to do this, but in practice did nothing: the terms of reference were not announced until June 1994, and the White Paper which resulted from the Review appeared in May 1995.[55] The delay in holding the Review is perhaps not surprising: nuclear power had proved politically difficult, and it was not clear who, apart from the two nuclear utilities, could have gained much from the Review process.

Part of the difficulty in getting the Review under way was that there appeared to be divisions within the Government machine about how to proceed, and in the end two parallel Reviews were conducted. The most wide-ranging Review was sponsored by the DTI, and its terms of reference were wide-ranging. The Department of the Environment held a separate review on the narrower but important issue of radioactive waste-management policy and a separate White Paper appeared on this subject in summer 1995.[56] There are important current waste-management issues (for instance NIREX plans for an underground 'Rock Characterization Facility' at the proposed site of its intermediate level waste repository went to an important Public Inquiry in late 1995). However, the Department of the Environment's White Paper, while clarifying long-term policy, produced no initiatives, and remaining attention is concentrated on the DTI's Review and White Paper.

The DTI White Paper covered three main subject areas: the case for new nuclear investment; the case for privatization, plus changes to existing subsidy regimes; and the management of nuclear waste and its associated liabilities.[57]

More than half of the White Paper is taken up with the first of these subject areas. The analysis is detailed and convincing. For the first time in the history of nuclear power in Britain, an official document is dismissive about the financial case for nuclear investment. The Government first argues that nuclear investment is extremely unlikely to take place in a market-led context.[58] While there are a number of specific reasons for this, the most important influence by far is the relatively high cost of capital that any private-sector investment needs to take into account. This is crippling to as capital-intensive a technology as nuclear power. The White Paper therefore concludes that only with significant Government support could new nuclear investment take place. It then considers the arguments in favour of such Government support. The possible arguments considered are: to help meet CO_2 reduction targets; to encourage diversity; and to gain wider economic benefits. After detailed examination of the arguments, all three are rejected.

On carbon dioxide reduction, the Government concludes that new nuclear power capacity would not be needed to help meet carbon emission targets until after the year 2010,[59] and conspicuously side-steps the question of whether nuclear would be a cost-effective way of achieving further CO_2 reductions at that point. On diversity, the Government argues that the market is likely to provide its own appropriate level of diversity, and concludes that in any case the analytical approaches to optimizing diversity are imperfect.[60] Finally, Government dismisses the view that the nuclear industry deserves any level of state support in preference to the claims of many other industries, none of which is similarly favoured. New nuclear investment is therefore consigned to the market place, where, as is now manifestly clear, there will be none, in the face of nuclear generating costs substantially higher than available from gas-fired generation, and risks that are also considerably larger.

It is in the second main area of the White Paper, privatization, that the greatest surprise emerged. This was a firm proposal to privatize the AGRs plus Sizewell B,[61] a prospect that had seemed unlikely even a few weeks before publication. The proposed strategy was to retain Magnox stations and their liabilities in the public sector – in a company to be called Magnox Electric, to be transferred in time to BNFL – and to privatize the remaining stations in a single company, British Energy, which would combine the two Scottish AGRs with the five English AGRs plus Sizewell B. This was in opposition to the advice given by OFFER,[62] which had suggested the more competitive solution of transferring two English AGRs to the Scottish company and one Scottish AGR to the English company, thus increasing competition both south and north of the border. The proposed solution did little for competition, given that the Magnox stations are too expensive to compete in any real sense.

At first the Government claimed that the FFL for nuclear would cease at the point of nuclear privatization,[63] and that this would lead to an 8 or 9 per cent fall in electricity prices at that point. However, it subsequently became clear that Nuclear Electric would be entitled to levy arrears that the company had voluntarily foregone in earlier years to avoid raising the value of the levy above 10 per cent. This would mean a smaller reduction in the levy, later in 1996, probably to around 3.5 to 4.5 per cent[64] instead of 1 or 2 per cent. While this was constitutionally possible, the revised proposals overlooked the fact that Nuclear Electric had already (early 1996) received more proceeds from the levy than it could ever spend before privatization. The collection of arrears from consumers could only serve to help the Treasury therefore, rather than nuclear power, as in practice all marginal levy receipts would flow directly to Government-held bank accounts or stocks. However, it remained the case that removal of the levy was a necessary condition for privatizing British Energy – otherwise the playing field for competition with other private generators could not be level.

Whether the new private-sector company can help in the competitive process by cutting costs significantly or improving efficiency is open to some doubt. The question of course is not whether a private company can cut costs – no doubt it will have the incentive and capacity to do so – but whether or not it can do so to a significantly greater extent than a company that remained in the public sector. It will be virtually impossible to improve the availability of nuclear stations significantly, mainly because unplanned shutdowns remain high and on-load refuelling is still not properly established. The labour force has already been substantially cut (and would in any case have been cut further); and favourable new contracts are already in place with BNFL. It does not seem likely that many savings in the future can be achieved through privatization as such.

The main benefit that would be likely to flow from privatization would be the commercial freedom that it would give the new company to invest in whatever it liked. Existing public-sector rules – reinforced by the presence of the large subsidy inherent in the FFL – compel the public-sector nuclear generators to invest only in nuclear technology. Given that NE and SN currently supply 27 per cent of Britain's electricity, it is plainly absurd to restrict them to investing in a technology that is commercially out of the question. Giving the new company commercial freedom is not itself a direct consequence of privatization, though privatization would provide a convenient opportunity to achieve it. This commercial freedom would have the effect of divorcing the future of nuclear technology from the future of the nuclear companies, and would almost certainly lead to new investment activity on the part of the new company. While seeking diversification, British Energy has for the moment explicitly ruled out early investment in any new generating technology in the UK.[65]

The feasibility of the privatization is far from certain. The Government hopes to raise £2.5–£3 bn from the sale.[66] However, the Government is asking the private sector to take on (at 2 per cent discount rate) some £7 bn of liabilities and, even on discounting assumptions which reduce the present value of these liabilities,[67] it is difficult to see circumstances in which the value of the new company to private investors could be much above zero. An independent study suggests a maximum value on net present value grounds of just over £800 m, with a probable value lower than this.[68] But all privatizations have proceeded with a discount on net present values, and the discount this time would have to be much larger than in earlier privatizations because of the extent and uncertainty of the liabilities that private owners would inherit with the AGRs/PWR, and because of the much larger political and general economic risk that accompanies nuclear power. If the Government were to achieve anywhere near £2.5 bn it would almost certainly only be by exempting British Energy from inheriting some of the liabilities associated with the AGRs. In practice, much the most likely exemption that might be gained by British Energy would be from the £4 bn worth of reprocessing bills that BNFL will charge for AGR fuel discharged from reactors to date.[69] Such an exemption might persuade investors to pay the £3 bn or so that the Government hopes to raise from the sale, but it needs to be made clear that consumers have already paid £3.5 bn to NE and SN for the reprocessing of this fuel,[70] and taxpayers would need to get all of this money back if they were not to start paying again for liabilities they have already now substantially paid for twice over (up to 1990 via the CEGB, and since 1990 via the levy).

The other problem to be faced over the privatization is the management of the even larger Magnox liabilities that will be left in the public sector. These amount to £17 bn (undiscounted) and, on the Government's estimate, to £8.5 bn discounted on the assumption that many liabilities need not be discharged till a century and more into the future.[71] Neither cash nor other liquid assets exist to help meet these huge liabilities, despite charges on consumers – including the fossil fuel levy – that have been exacted since 1976. Future Magnox income is optimistically valued by Government at a net present value of £1 bn,[72] leaving virtually all Magnox liabilities to be paid for by future consumers and/or taxpayers. Whether such a formulation, together with a probable low sale value and/or large Government guarantees, is politically acceptable is not yet clear.

CONCLUSIONS

Privatization of the ESI proved to be the occasion of major upheaval in the nuclear power sector. From having been perceived by Government as relatively cheap, politically reliable and potentially privatizable, nuclear power came to be seen in 1989 as expensive and financially risky, politically difficult, and impossible to privatize. The botched attempt to privatize nuclear power in 1989 illustrated some of the problems of a politically tight overall timetable for privatization, and also some of the difficulties resulting from an attempt to change virtually all of the elements of the old system at one time.

The new public sector companies faced major problems in re-building the financial and economic credibility of nuclear power, but they have succeeded to an extent that would have been hard to foresee. Such was the success of this operation – raising AGR output to almost unimagined levels and controlling costs effectively – that by 1995 a more limited privatization could again be contemplated. The success of this effort is not certain at this stage, though the Government will clearly be reluctant to preside over a second unsuccessful attempt at nuclear privatization.

If privatization of the AGRs and Sizewell B does work, it will leave behind a very large future liabilities burden from the Magnox stations, with a very limited future public-sector nuclear income to set against these costs. Consumers and taxpayers will face very large bills for reprocessing, decommissioning and waste management over a very long future period. The management and funding of long-term nuclear liabilities will remain a major public issue whatever the result of the privatization.

The privatization process proved very effective at exposing the high costs and financial risks of nuclear power in British conditions. The transparency in nuclear costs which the separation of nuclear power has produced has been valuable, but it has also exposed the clear fact that in the privatized British electricity system, with its high cost of capital, new nuclear investment is hopelessly uneconomic. This has been explicitly recognized for the first time by Government, and there has also been official rejection of further subsidy for nuclear construction. There will clearly be no nuclear investment in Britain for some years, and only if Government becomes much more interventionist in the environmental or diversity areas is the case for nuclear investment ever likely to be re-opened, let alone settled in favour of nuclear power.

8

Renewable Generation – Success Story?

Catherine Mitchell

Renewable electricity generation was supported by a market enablement programme for the first time in the UK as part of privatization. Renewable energy projects were able to obtain a premium price per kWh of generation if they were successful in their application for a contract under the non-fossil fuel obligation (NFFO). The subsidy is the difference between the average monthly Pool price and the premium price. The NFFO obliged the regional electricity companies (RECs) to take a certain amount of nuclear and renewable electricity. The renewable NFFO was to be made up of five Orders or tranches of contracts between 1990 and 1998. Together these Orders were intended to fulfil the Government's policy of working towards 600 MW declared net capacity (DNC), later revised first to 1000 MW DNC and then 1500 MW DNC, of new renewable capacity by 2000. This target is essentially an arbitrary and political figure, and it is the major determinant of the subsidy paid over time.

In late 1994 the Government announced the successful contractors of the third non-fossil fuel obligation (NFFO3) and in late 1995 announced the application procedures and rules of the fourth NFFO (NFFO4), due to be finalized in 1997. Another NFFO Order is due to be set in 1998 (NFFO5). There is no stated policy of support for renewable energy in the UK beyond NFFO5 and Government expenditure on renewable energy research and development is due to be phased out by 2005, although a fundamental policy review is promised for 2000.

The chapter analyses what impact privatization had on renewable energy deployment in the UK. It discusses UK policy towards renewables prior to privatization; to what extent support for renewable energy was an intended policy of privatization; the NFFO process and its results so far; and some key issues for the future. A concluding section discusses, among other things, whether renewables could or would have been supported in the UK without privatization.

UK Renewables Policy

Current Policy

In July 1993, the Minister for Energy (subsumed within the Department of Trade and Industry following the 1992 General Election) clarified the Government's policy on renewable energy.[1] He stated that:

> Government policy is to stimulate the development of new and renewable energy technologies where they have the prospect of being economically attractive and environmentally acceptable in order to contribute to:
>
> - diverse, secure and sustainable energy supplies
> - reduction in the emission of pollutants
> - encouragement of internationally competitive renewable industries.

He explained that:

> the purpose of the NFFO Orders is to create an initial market so that in the not too distant future the most promising renewables can compete without financial support. This requires a steady convergence under successive Orders between the price paid under the NFFO and the market price. This will only be achieved if there is competition in the allocation of NFFO contracts.

The Minister's statement was recycled in a number of Government publications which confirmed Government support of market enablement via the NFFO or research, development and demonstration (R,D&D) for solar, onshore wind, wastes, hydro, energy crops, photovoltaics and fuel cells. They also stated that wave, geothermal, tidal and offshore wind are classified as unlikely to contribute substantially to UK energy supply in the foreseeable future and essentially have had their funding cut; and that the

Government is working towards 1500 MW DNC of renewables generating capacity by 2000.

However, the most recent Government statement concerning renewable energy, the announcement of the fourth Order, justified the policy on grounds of stimulating convergence between electricity prices under the NFFO and the market price for electricity. It omitted any mention of reduced emissions or the encouragement of competitive industries. The aims and commitment behind Government policy are therefore somewhat opaque.

RENEWABLES POLICY BEFORE PRIVATIZATION

The UK renewable energy policy prior to privatization was based on research and development (R&D) programmes and, latterly, a few demonstration projects. The Government had undertaken a renewable energy research programme since the mid-1970s. In 1982, the Government Advisory Council on Research and Development for fuel and power (ACORD) undertook a major review to establish a methodology by which the potential and cost of a technology could be assessed. Technologies were assessed as strongly placed; economically good; promising; or long shots. As a result of their placing under this methodology, renewable energy technologies were either supported or their funding curtailed, as occurred with wave power. The 1982 ACORD Report was updated in 1986. A new renewable energy review was then published in 1988 (Energy Paper 55), which has recently been overtaken by another renewable energy review in 1994 (Energy Paper 62). This confirms that the renewable energy R&D programme will extend to 2005, although it makes it clear that its continuation to that date will depend on the outcome of the next fundamental review in 2000. Energy Paper 62 envisages that declining Government R&D support will be offset, and even bettered, by industrial and other contributions.

At the time of the 1973/74 oil shock, there was virtually no Government support for renewables – in strong contrast to the large Government support for nuclear power at all stages of the nuclear fuel cycle, including R&D. Total UK R&D expenditure on renewables from 1977 to 1994 was £215 m in 1994 prices, with average annual expenditure around £12 m in 1994 prices.

The largest percentage expenditure was on wave power, followed by wind, geothermal and solar technologies. From 1978, the percentage given to wind energy gradually increased to about 30–40 per cent of the programme. Spending on wave power, which had been 60 per cent of the programme expenditure in 1978, fell to 10 per cent in 1991. Energy Paper 62 placed wave power in the 'watching brief' category which essentially means no more funding. Geothermal funding rose for a decade, becoming of equal importance with wind energy between 1981 and 1984, and then gradually declined. Tidal and solar have received small amounts throughout the period; the biomass programme is gradually increasing with photovoltaics (pv) receiving £60,000 of funding for the first time in 1991–92. Thus, there have been large changes in government priorities for the different types of renewables over the past twenty years which, although no doubt unavoidable to some extent, have reduced the continuity of the development effort.

When comparing the technologies which received R&D spending to those supported by the NFFO we can see that there is limited overlap. The three NFFOs supported medium-size wind turbines (ie turbines of 300–750 kW), landfill gas, sewage gas, hydro, biomass gasification and waste to energy plants while R&D mainly supported large-scale wind turbines (3 MW), landfill gas, hydro, geothermal, wave and tidal. The main overlap was with landfill gas and hydro and smaller-scale wind turbines.

The major proportion of the total R,D&D expenditure was on technologies which have, to all intents and purposes, been curtailed (wave and geothermal), or changed tracks, such as the wind programme. A quarter of the wind programme was spent on the 3 MW Orkney wind machine with a view to developing of large-scale turbines. This turbine is due to be dismantled and the wind programme has now radically altered to investigate smaller-scale turbines. A major UK re-assessment of wave energy occurred in 1993 where the potential of wave power was downgraded and the author of the report moved to another department.

It was against this background that the National Audit Office (NAO) undertook its inquiry into 'the effectiveness of the DoEn's overall approach to meeting their objectives through planning, evaluating and managing its renewable energy R,D&D programme'.[2] The NAO examined three technologies in particular: landfill gas, hot dry rocks (HDR) and wind. Their general conclusions were: that the DEn's methodology of choice of technology support was soundly based, but the programme had not provided sufficient support for projects aimed at the export market (estimated worldwide as £31 bn by the EC), although this was one of the stated aims of the programme; and 'the earlier influence of the main customers' (ie the CEGB) led to a few projects and programmes receiving a large share of the total budget. The CEGB 'was mostly interested in developments capable of bulk energy generation' such as large wind turbines, tidal power and HDR. Such R,D&D programmes were expensive and one third of the total available funds were consumed by the Severn and Mersey Barrages, large and vertical-axis wind turbines and HDR, none of which have come to fruition. This exemplifies a problem of the R&D programme, highlighted in the NAO Report, namely that it has supported one technology over, and to the detriment of, others.

In practice, development of renewable energy technologies in the UK prior to the NFFO, other than for hydropower in Scotland and pumped storage in Wales, was very limited. The majority of around 200 hydro plants in the UK were generally smaller than 300 kW installed, were self-financed by individual, private owners and often attached to attractive mill-houses, as they often had been for several centuries. Investment in anaerobic biogas (for example, farm or abattoir waste) plants had been undertaken by about 20 self-financing individuals or small companies. Investment in sewage gas plant was primarily being undertaken by water companies as part of an ongoing programme of water privatization to reduce sewage dumping in the sea, which happily coincided with the NFFO. There were 32 existing landfill gas sites prior to the NFFO, all undertaken and internally financed by the operating companies of the landfill sites, and generally used for heat rather than electricity. Only four municipal and general waste mass-burn incinerators with energy recovery units existed, all paid for by the local authority in question: North London Waste Authority plant at Edmonton, Nottingham, Sheffield and Coventry. No specialist waste-to-energy plant (such as for hospital waste or tyres) existed. Private investment in wind power consisted of half a dozen single turbines of 65–95 kW DNC, a few turbines of around 5 kW and several 25–500 W turbines for individual light-bulb-type output on boats, caravans and so on. In addition to this, there were ten R,D&D turbines rated between 130kW and 3000kW installed at six sites in the UK. There were no commercial or demonstration tidal, wave or offshore wind projects, although a wave project began on the Isle of Islay in Scotland in 1990.

These renewable energy schemes in existence prior to the NFFO were able to sell their electricity, if they were connected to the grid, under 1983 Energy Act terms. Under the Act, electricity boards were obliged to buy electricity from independent generators, but they paid the renewable generators an average of 30 per cent less than the CEGB for their electricity.

THE NON-FOSSIL FUEL OBLIGATION

The NFFO requires the RECs to buy a certain amount of nuclear and renewable electricity. The RECs pay the generators a premium price for the renewable electricity and the difference between the premium price and the average monthly pool purchasing price (PPP) is reimbursed to the REC by the Non-Fossil Purchasing Agency (NFPA) from the fossil fuel levy (FFL). This levy applies to all electricity generated by fossil fuels and is added to consumers' electricity bills.

By late-1995 there had been three NFFO Orders. The first in 1990 (NFFO1) provided contracts for 152 MW DNC of landfill gas, sewage gas, hydro, wind energy, waste-to-energy and biomass projects. The second in 1991 (NFFO2) provided contracts for 472 MW DNC for projects based on similar technologies. The third Order of 627 MW DNC was awarded in December 1994 (NFFO3) and included biomass gasification for the first time but excluded sewage gas. It is envisaged that both NFFO4 and NFFO5 will contract for 400–500 MW DNC, although only about two-thirds is expected to come on line. The renewable premium prices have been paid for by 1–5 per cent of the FFL, which has been set at between 10 and 11 per cent of the electricity price since 1990.

The NFFO provides a contract for the supply of electricity to the RECs; it provides a subsidy to the contractors and it accepts the principle that paying a premium price for electricity from near-market technologies is an appropriate means of assisting those technologies to competitiveness. These provisions are important because not only do the RECs have no obligation beyond the NFFO to buy renewable electricity (the terms of the Energy Act having lapsed), but they are under an obligation through their Licence to operate to undertake 'economic purchasing' to ensure that their customers are not paying more than is necessary for their electricity.

A number of factors explain the increased support for renewables, including the eruption of concern over climate change towards the end of the 1980s, the low price of fuel which meant that the Government could no longer argue that renewables could develop without a subsidy, but chiefly because the NFFO and the fossil fuel levy were an acute embarrassment to the Government which had failed to privatize nuclear power.

The renewable NFFO developed out of the need to find a means of supporting nuclear power, once it was realized that the nuclear portion of the electricity supply industry (ESI) could not be privatized in 1989. The Competition directive required that the Government obtain permission from the European Commission (EC) for a levy to pay for nuclear power. The Government asked the Commission to accept a levy to pay for non-fossil generation, specifically not mentioning nuclear power. The Commission agreed to a levy but only until 1998.

No mention of support for renewables appeared in the privatization literature until the announcement of the levy. When the levy was first announced it did not set a capacity of renewable energy to be supported through the renewable NFFO. It was only when the timetable for the privatization process began to slip that renewables and nuclear power were separated and the Government announced that the renewables NFFO would support 600 MW DNC.

The renewable NFFO was justified by the Government on two main grounds: it would help new technologies into the market place; and it would increase the number of independent power producers (IPPs), an aim of electricity privatization. Part of the Government rationale for the NFFO approach was that it avoided the constraints of public expenditure, since it was not originally counted within the public sector borrowing requirement (PSBR). The subsidies were paid for by consumers, not taxpayers.

The Government initially asked the EC for approval for a levy to support non-fossil electricity for an indeterminate period, expected to be at least 15 years. A compromise between the UK Department of Energy (DEn) and the Commission was reached whereby the levy would be set for eight years. The Commission later stated that it 'did not wish in 1990 to grant authorization for support of nuclear power beyond 1998'.[3] The probable reason for this is that the NFFO and FFL appeared to be incompatible with the aim of introducing full competition with the abolition of the tariff market franchise in April 1998. However, the Commission let it be known in 1990 that it would look favourably on a proposal by the UK to extend the NFFO for renewables beyond 1998. Such an exemption took until mid-1993 to propose and agree – just before the conditions of the NFFO3 were announced. Despite various justifications, renewables were supported in the privatization process as a result of, and linked to, nuclear power.

EXPERIENCE TO DATE

This section reviews UK experience of the use of the NFFO mechanism until early 1996 to stimulate renewables development and the changes and issues which have arisen. It briefly reviews each of the NFFO orders in turn.[4]

NFFO1

The first NFFO order was essentially cost-plus, with the subsidy aimed at giving sufficient economic rent to attract developers. Each project was assessed separately and no direct competition occurred between projects or technologies.

The initial period was one of some confusion. The regional electricity boards, which were vetting the cost-justification[5] proposals for the applicants, were at the same time grappling with the problems of their own transformation into RECs and of creating CCGT and other generation company subsidiaries. The renewables developers regarded the boards as potential competitors and disliked having to send them their own cost and financial details. The RECs, NFPA and OFFER were set up in late-1989 and early-1990 and there was initially confusion amongst them and the developers about their responsibilities. OFFER had to undertake a 'will-secure' test for each project. This included a consideration of the financial performance of the project, assurances on the availability of the site, the likelihood of obtaining the necessary planning consents, the technical viability and feasibility of the commissioning date, the soundness of the estimated capital and operating costs, and whether adequate funds were available.

As the number of applicants rose to 370 (far above expectations), DEn advised the RECs of the ceiling or 'cap' price the NFPA would be

Table 8.1 Status of 1990 NFFO projects

Technology band	*Number of schemes contracted*	*Number of schemes commissioned*	*Contracted capacity (MW DNC)*	*Capacity commisioned as of 30.9.95 (MW DNC)*
Wind	9	7	12.21	11.7
Hydro	26	21	11.85	10.00
Landfill gas	25	20	35.5	31.68
M&GIW	4	4	40.63	40.63
Other	4	4	45.48	45.48
Sewage gas	7	7	6.45	6.45
Totals	*75*	*63*	*152.12*	*145.94*

Source: DTI, *Renewable Energy Bulletin 6*, December 1995.

prepared to pay and asked host RECs to invite developers to resubmit their bids accordingly. When many of the new wind and hydro sites could not meet the 6p/kWh price cap, DEn raised the price to some wind generators by 50 per cent (to 9p/kWh). As a result of the confusion over the application process, the sudden introduction and consequences of the 1998 end-date and the mix-up over the price cap the number of applicants winnowed down to 100 in mid-1990, of which – as shown in Table 8.1 above – 75 received contracts to build 152 MW DNC.[6]

NFFO2

The main change under the second Order was that contracts were now awarded on the basis of competitive bidding in technology bands. Wind projects competed with wind projects, sewage-gas projects with sewage-gas projects and so on. Reminiscent of the electricity Pool price which is paid to each plant called up to despatch, irrespective of its bid prices, the 'strike' or marginal price of each technology band was paid to each contract within the technology band. The premium prices were particularly high since NFFO2 projects, still with a 1998 end-date, had less time to run than those under the first Order; 11p/kWh was paid for wind energy (around four times the average Pool price). The high 'strike' prices attracted not only many developers but also the attention of critics who argued that 'strike' prices merely gave low-price bidders a windfall gain.

Four main problems were encountered. Firstly, wind power ran into considerable criticism. Competition requires the bringing together of a number of projects at one time. This results in waves of development. As several wind farms began to be commissioned at the same time there were fears particularly in Wales of the wind farms' visual intrusion and land-take, and the apparent lack of coordination between the administration of the NFFO and local planning authorities.[7]

Secondly, small-scale projects and independent generators found it particularly hard to obtain contracts – small-scale projects because the costs were generally higher than for larger projects, and the independents because they found it hard to obtain finance. Only two projects within NFFO2 were developed by independent developers who did not have their own equity. All other projects initially developed by independent companies were forced to accept equity from RECs, generators or water companies at a high cost of capital, thereby reducing their own returns.

Table 8.2 Status of 1991 NFFO projects

Technology band	*Number of schemes contracted*	*No. of schemes commissioned*	*Contracted capacity (MW DNC)*	*Capacity commisioned as of 30.9.95 (MW DNC)*
Wind	49	26	84.43	53.91
Hydro	12	7	10.86	10.05
Landfill gas	28	26	48.45	46.39
M&GIW	10	2	271.48	31.5
Other	4	1	30.15	12.5
Sewage gas	19	19	26.86	26.86
Totals	*122*	*81*	*472.23*	*181.22*

Source: DTI, *Renewable Energy Bulletin 6*, December 1995.

Thirdly, the 1998 end-date caused such haste that contract holders had to buy foreign wind turbines. The only UK manufacturer, Wind Energy Group (WEG), was working with National Windpower and did not have the capacity to supply turbines to all the other contract holders within the necessarily short period of time available. As a result, of the 1990 and 1991 NFFO contracts, 345 (83 per cent) of the turbines were imported.[8]

Lastly, over 200 MW DNC of waste-to-energy projects were terminated because the 1998 end-date gave too little time for their economic development. It can be argued that this triggered the decision to ask the European Commission for exemption from the 1998 end-date for the third Order. The EC's concession on this was especially important for waste-to-energy projects since stringent emission controls were being introduced by HMIP under an EC directive on UK incinerator emissions, and many incinerator plants will have to be closed.[9]

Table 8.2 shows that 122 projects with a total capacity of 472 MW DNC were contracted under NFFO2, of which 49 were wind, 28 landfill gas, 19 sewage gas, 12 hydro and 10 waste combustion.

NFFO3

The main change under the third Order was that, although competitive bidding within technology bands continued, contractors were awarded

Table 8.3 1994 NFFO contracts

Technology band	*Contracted capacity (MW DNC)*	*Number of projects*	*Lowest contracted price/kWh*	*Weighted average price/kWh*	*Highest contracted price/kWh*
Wind (exceeding 1.6 MW DNC)	145.92	31	3.98	4.32	4.8
Wind (below 1.6 MW DNC)	19.71	24	4.49	5.29	5.99
Hydro	14.48	15	4.25	4.46	4.85
Landfill Gas	82.07	42	3.29	3.76	4.00
Municipal and industrial waste	241.87	20	3.48	3.84	4.00
Energy crops & agricultural & forestry waste					
Gasification	19.06	3	8.49	8.65	8.75
Residual (other)	103.8	16	4.9	5.07	5.23
Total	*626.91*	*151*	–	*4.35*	–

Source: DTI Press Release (1994), *Wardle Makes Third Renewable Energy Order*, 20 December.

their bid prices, as opposed to the 'strike' price which was a feature of NFFO3. The third Order therefore moved towards reducing the economic rents available from the contracts.

Table 8.3 shows the size of the Order and the spread of prices paid

Table 8.4 NFFO price falls

Technology	*Band price NFFO2 price/kWh*	*Band price NFFO3 price/kWh (average)*
Wind	11.0	4.32 (1.6 MW DNC +)
	–	5.29 (under 1.6 MW DNC)
Hydro	6.0	4.46
Landfill gas	5.7	3.76
Waste Combustion	6.6	3.84
Other Combustion	5.9	5.07
Sewage gas	5.9	–
Average	*7.2*	*4.35*

under each technology band, while Table 8.4 shows the resulting price reductions under NFFO3 compared with NFFO2. The price reductions were due to several factors. Fixed costs were substantially reduced by the fact that contracts were now for 15 years rather than the 8 or fewer years of the two previous Orders.

The generators were awarded their bid price rather than the previous strike or marginal price. The prices charged by planners, lawyers and other professionals involved in project development fell as they gained experience. Prices of renewables hardware also fell significantly. As a result the cost per kW of installed wind energy capacity fell from £1000 under NFFO2 to £700–750 under NFFO3. This was partly due to the technical improvements associated with the move from 300–400 kW turbines in the NFFO1 to 400–750 kW machines in the NFFO3. The prices for other renewable technologies also fell, but not so much. Whereas many NFFO2 projects were financed externally, it is expected that many NFFO3 projects will be financed from internal sources, giving a lower cost of capital. The fact that contracts may be taken up within five years of the award date gives ample time to seek and obtain planning permission (particularly important for waste-to-energy projects, many of which fell by the wayside in the NFFO2) and makes it easier for projects to incorporate British hardware, now a stated Government policy.

A new problem was that NFFO3 was heavily oversubscribed – 141 projects, comprising 627 MW DNC, were awarded contracts, but 380 projects totalling 1870 MW DNC were refused. This led to calls for more clarity about the next Order – its size and minimum prices, and so on – and possible alternative forms of support for developers with projects they feel are worth developing but who think they are unlikely to obtain a contract under competitive bidding.

Another significant change was that two new clauses were inserted in NFFO3 at the behest of the RECs. A 'levy-out' clause stated that if the levy ceased during the contract, the RECs would not be liable for making up the difference between the Pool price and the premium price. A 'supply-out' clause stated that if renewable energy generation exceeded 25 per cent of the RECs' supply business, the RECs would not have to take the renewable electricity. Although these two clauses therefore placed the risk of developing renewables firmly with the developers, the terms of NFFO3 contracts overcame many of the problems of the previous Orders and also highlighted the potential supply of renewable electricity stimulated by the support mechanism.

Lastly, waste-to-energy projects were again awarded a large proportion of the contracted capacity, going some way to making up for the problems of NFFO2. Nevertheless, most of the 20 projects under NFFO3 have neither planning permission nor their fuel supply guaranteed, and competitive bidding means that the technology used is the cheapest rather than the least environmentally damaging – for example waste combustion solely for power generation as opposed to combined heat and power, or waste recycling outside the combustion process. Friends of the Earth consequently condemned the contracts and have argued that NFFO's should be part of an integrated waste management strategy.[10]

NFFO4

In late 1994, the Government announced the fourth Order and said that 400–500 MW DNC would be contracted in both NFFO4 and NFFO5 and that the terms would be similar to those of NFFO3. The technologies which continue to have government support are wind energy, hydro, landfill gas, electricity from energy crops and forestry waste using gasification. Biomass technologies based on steam generation which were supported in NFFO3 are henceforth to be excluded. Electricity from agricultural waste and food processing, based on anaerobic digestion received a technology band for the first time. Municipal and industrial waste continued to be supported, but this time split into two bands so that generating plants incorporating combined heat and power also qualify. Whether by accident or design, this time no attempt was made to justify the Order on environmental or diversity grounds. If deliberate, this omission may signal the end of Government support for renewables after NFFO5 in 1998.

As NFFO3, the fourth Order was heavily oversubscribed. In a Press Notice,[11] the junior Energy Minister Richard Page announced that bids for nearly 900 projects had been received from potential generators with a total capacity of 8,387 MW DNC. Of this, nearly 6 GW came from potential waste to energy plants. The awarding of the contracts is expected early in 1997 and will inevitably disappoint the majority of bidders.

Key Issues

After NFFO5 the Government will have to decide whether consumers should continue to subsidize the development of renewable generating capacity and, if so, on what scale, whether taxpayers or consumers should pay for further R&D on renewables and, if so, which technologies should qualify. The judgements involved have been difficult from the outset, partly due to the lack of cost and performance data (which was part of the reason for launching the pilot renewables programmes) and partly due to the uncertainties surrounding the future prices of conventional forms of energy. We now consider four key issues which require resolution if renewables are to have a chance of developing beyond NFFO5.

The Convergence Price

The first issue is whether the convergence price – the UK Government's goal for renewable energy prices and the price which host RECs would have to pay for renewables' electricity if the RECs (as opposed to industrial and commercial users) bought it – should be based on money costs alone or whether it should include estimated benefits for reduced environmental impacts and for embedded generation. As with any electricity generating technology, valuing the environmental benefits of renewable energy is difficult – particularly when considering the opposition to incinerator plants and certain wind farms.

EMBEDDED GENERATION

It is often claimed that small-scale, localized and dispersed renewables projects offer the advantage of 'embedded' generation within the distribution system of the host REC which avoids a number of costs associated with large, centralized generation, generally to do with transmission and distribution (electrical losses, capital and maintenance costs, security of supply and operational constraints). It can therefore be argued that embedded generation should benefit from these avoided costs, either through a payment intended to represent those benefits or as a result of establishing the value of electricity at the point of use. But the marginal or avoidable cost of transmission is intrinsically difficult to estimate not only for the current system but also because of the problems of estimating future technical change (eg superconductivity) and of predicting future changes in the geographical pattern of generation and how it might be affected by the trend towards more cost-reflective transmission and distribution charges. Without either an environmental or a locational credit and with more expensive financing costs, the majority of renewable energy projects will be uncompetitive against most other forms of electricity: if the Government intends to support renewable energy beyond NFFO5, some softening of the competitive approach must occur. An obligation to buy renewable electricity at a price which includes credits, analogous to the standard payment used in most European countries and discussed below, may be the minimum support required to ensure a continuing UK renewable energy industry.[12]

THE RETAIL MARKET

Another key issue is the extent to which renewables can enter the retail market in 1998. Since the RECs will still have a monopoly of their distribution system, renewable generators and buyers will have to negotiate with them for the use of the grid, for back-up and so on. The extent to which the regulator enforces rulings that provide fair and transparent prices for use of the system and for fair back-up prices and removal of obstacles to connection to the grid will all affect the type and amount of support renewables need. Nevertheless, this market is likely to be seen by the DTI as offering the most potential for renewables if no other forms of support are available to them post-NFFO5.

Environmental Planning Approval

A fourth issue is the relationship between economic regulation of renewables projects which rests with OFFER, and environmental planning approvals which rest with local planning authorities. As in the water industry, this relationship has not been as effective as it should have been. Numerous applications have fallen at the environmental planning stage because too much time elapsed to complete construction and leave enough time to recover the full costs of project development. OFFER assesses the project application and applies a number of tests of the feasibility and viability of the proposed project (the 'will-secure test') and the developer is left to use their 'best endeavours' to obtain planning permission. For their part, the local planning authorities appear to have had inadequate guidance early on. Their duty is to uphold planning laws and principles but also taking account of relevant government policy which, in this case, was the subsidized development of renewables energy capacity based on competition, the opposite of a 'planned' support mechanism.

The issue for local planning authorities has been that, because building up renewables development has been a government policy since electricity privatization, where should they draw the line if they are presented with applications on a scale which would take up many of the best sites in a given area? This is compounded if the particular area affected is of considerable aesthetic or environmental value and if other parts of the country receive only very few applications. The final target of the NFFO is 1500 MW DNC of new renewable capacity by 2000, roughly the equivalent of 2000 MW installed and about 3 per cent of the electricity supply. It is clear to all involved that if this capacity is to be successfully commissioned without another major environmental confrontation, as occurred in 1993–1994 concerning wind energy, widely approved planning laws need to be established.

The Comparative Cost

How does the British method of supporting renewables – in tranches under successive NFFO Orders and subject to competition – compare with the support methods in Denmark, Germany and the Netherlands, all of which pay a standard payment per kWh of renewable electricity?

Denmark, along with the United States, has been supporting wind energy for a long time. Early on there were capital grants and generous tax incentives. In 1985, the Danish Government agreed that the electric utilities would be the primary developers of wind power yet the majority of developments continue to be undertaken by single-family or cooperative developers. The utilities pay 85 per cent of the small-consumer price for every kWh of wind electricity generated and a different percentage for other types of renewable electricity. In addition, the energy tax, the CO_2 tax and VAT on the CO_2 tax are all refunded to renewable generators. Non-utility investors have recently been freed from restrictions placed on them concerning renewables development, and have obtained large finance for this type of investment from small banks and building societies. The Danish support system is simple yet flexible and does not rely on competition.

Germany's support system is more complex. A standard payment of 90 per cent of the small-consumer price is available for all wind electricity generators. Payments were also available from the Research and Technology Ministry (BMFT) until January 1996 and continue to be available from some regional governments. Eligibility for all these schemes was fairly open, but applicants did not receive support from one source if they were already in receipt of payments from another source. Liberal financing of renewables projects is also available from numerous banks, including a cheap interest-rate scheme provided through the Ausgleichsbank. The standard payment appears likely to continue but continued support for the extra measures is uncertain. The electric utilities complain of a high cost of supporting renewables and wish to limit their responsibility for providing and paying for grid connections.

The Netherlands established a system of support in 1990 which combined subsidies with utility action. However, this policy was judged to be unsuccessful in increasing the rate of deployment of renewables and a new set of support mechanisms were introduced in January 1996. These shifted the policy emphasis from subsidies towards tax and fiscal incentives.

In 1990, the main renewables support mechanism was a system of capital grants of up to 35 per cent of total capital costs, which were paid from general tax revenues through NOVEM, the Environment Agency. Applications were made once a year, without competition or a required rate of return, and vetting procedure was straightforward. Demand, however, was much greater than NOVEM's budget could support. This mattered less to the extent that there was a second strand of support consisting of a 2 per cent levy on the price of electricity for small consumers. Although the utilities used most of this to support CO2 reduction measures of their own choice, in particular CHP, some utilities provided an additional premium payment for renewable energy or developed their own projects. Thus, Dutch utilities chose whether and how to support renewables, especially projects which did not obtain a NOVEM grant.

In order to reduce the extent of utility control on renewable deployment a number of new mechanisms were introduced in 1996. Subsidies are no longer available from NOVEM but the MAP payments continue so that utilities can support renewables if they wish. However, green funds have been established to provide investment incentives; certain environmental investments are also eligible for accelerated depreciation benefits; renewable energy equipment has been transferred to a green (low) VAT bracket; and finally the ecotax (first levied in January 1996) is refunded to renewable electricity generators. As a result of this, wind electricity now receives a standard tariff throughout the Netherlands made up from an avoided cost payment, the MAP payment and the ecotax refund. Both types of support mechanisms, standard tariffs versus premium payments based on competitive bidding, have advantages and disadvantages. For the continental method of standard payments, the advantages are that developers can apply at any time, they know what payment they will get, there is no hurdle rate-of-return requirement, and the vetting is simple. The disadvantage is the difficulty of setting the subsidy without resulting in either substantial economic rents for lower-cost projects or insufficient inducement to develop higher-cost projects. The advantages of the British system are that the concept of price convergence and competition within technology bands are more likely to encourage economic efficiency with the ultimate aim of renewable energy projects which are competitive with conventional energy technologies. Provided that the Government makes the right choices (which on past experience cannot be assumed), it usefully allows the Government to choose which renewable technologies to support, rather than spreading resources thinly over all renewable technologies. The disadvantages of the British system are that the tranches result in very busy periods followed by lulls, it is difficult to coordinate with planning requirements, the administration is relatively complex and is yet another task for OFFER, whose primary task is economic regulation of the privatized electricity industry.

The comparative costs of the two basic types of support mechanism can be estimated only approximately due to problems of exchange rates, purchasing power, and the indirect support mechanisms in Germany. We made comparisons in terms of purchasing power standards (PPSs) per MWh, where PPSs are a measure which combines exchange rates with a country's purchasing power. The subsidy in each country was calculated by subtracting the Pool price from the premium price paid to renewables. This was multiplied by the annual output and number of years to obtain the total subsidized output, which was annuitized and converted to PPSs. The result was then divided by the total output to obtain a PPS/MWh. Separate calculations were done to estimate the cost of generating a kWh of electricity from the same turbine and the project rate of return, assuming the same project were undertaken in each of the countries. The difference in cost is mainly due to differences in the cost per kW installed, the cost of capital, contract length, and level of subsidy.

Table 8.5 suggests that NFFO2 resulted in a cost of electricity, a level of subsidy and an average rate of return on renewable projects in England and Wales which were substantially higher than was the case in the continental countries. By NFFO3, considerable progress had been made in reducing the cost of electricity, the rate of return and particularly the level of subsidy to levels comparable with or better than those obtaining on the

Table 8.5 The cost of subsidies for wind energy: the net present cost of national mechanisms based on UK costs assumptions

		Cost of subsidy (PPS/MWh)	*Rate of return (%)*	*Cost of electricity (PPS/MWh)*
Denmark		20	15	48
Germany		24	11	56
The Netherlands	(1)	11	21	13
	(2)	29	25	32
UK (1991 NFFO)		81	30	97
(1994 NFFO)		11	13	54

Source: C Mitchell (1995) 'A Comparison of the Means and Cost of Subsidising Wind Energy', Part A: *Journal of Power and Energy*, Proc Instn Mech Engrs, vol 209, p 185–188.
1 Represents maximum grant (ie 35% national investment grant and 25% MAP payment) but no subsidy per kWh.
2 25% MAP grant and maximum utility payment taken from E van Zuylen and A van Wijk (1993) *Tarieven en Subsidiebeleid Voor Windenergie in Nederland en Het Buitenland*, Ecofys, Utrecht, The Netherlands.

continent. The estimates therefore seem to confirm the view that, after an initial period of unduly high rewards for developers which succeeded in attracting numerous new entrants, competition and convergence succeeded in reducing the price, the rate of return and the subsidy.

CONCLUSION

Support for renewables through the NFFO mechanism, which was designed in 1989/90 to bolster nuclear power when it could not be privatized, has proved successful in stimulating the growth of a nascent renewables industry in Britain.

Privatization as such was not strictly necessary for this purpose, since other forms of subsidy could have been provided and made to work equally well, given the right terms and conditions. Beyond that, it can only be hypothesized whether similar support would have been provided in the absence of privatization. If the NFFO mechanism was used to promote renewables solely in order to reduce embarrassment over the real purpose of the NFFO (to support nuclear power, and in particular to provide for the very large liabilities for fuel reprocessing and decommissioning), it would be reasonable to suppose that comparable assistance for the commercialization of renewables would not have been provided had public ownership continued. But this ignores that government commitments were made in the context of the 1992 Rio Conference which focused particularly on the problem of global warming. Had electricity privatization not taken place and had the NFFO mechanism not been available, the various national commitments made in the context of the Rio Conference might have persuaded the Government to provide some other effective form of support for renewables. It would have been difficult for the Government to have played its chosen role in the international diplomacy of global warming without a renewables 'showcase' back home.

There are considerable problems with international comparisons of different subsidy mechanisms and it is difficult to draw any overall conclusion about which mechanism is the best. The continental approach of standard payments is undoubtedly easier to administer and provides more certainty – an important factor in attracting investors – and it is easy to include credits for renewables in such a system. The British system has had two advantages. Firstly, the correction of two early mistakes, with the lengthening of the period for which the subsidized price was available and the change from the 'strike' price to paying bid prices, have substantially reduced the subsidy element – which was initially undoubtedly high. Secondly, the whole approach has been based on the idea of a price-convergence path (meaning that the subsidy should taper and then cease), which is in line with the idea that infant industries should be assisted, but encouraged to become competitive. The administration of the renewables NFFO scheme has had these desirable features and the results have been appreciable – especially compared with the period 1978–1989 when Government support for renewables was predominantly for preliminary studies and pilot schemes, but put no significant money towards development and prototypes nor commercialization for those renewables technologies where it was most needed.

Since 1986, the main factor working against both renewables and nuclear power has been cheap natural gas. Since 1990 the 'dash for gas' has dominated electricity generation, based on cheap gas and combined-cycle gas turbine plant which are quick and cheap to build, efficient in operation, and effective in reducing atmospheric emissions (although not ultimately as effective as renewables sources, assuming these had been developed on a scale comparable with the 'dash for gas'). Due to the promotion of competition in UK gas markets, the 'beach' (wholesale) price of gas is now volatile, the short-term 'marker' price fell from 18–20p/therm to around 9p/therm during the course of 1995, and future price levels are exceptionally uncertain. If gas costs remain as low as they are in early 1996 it will be more difficult for any Government to maintain effective market support for renewables beyond NFFO5. If the reverse occurs, and the pressure of rising demand on finite resources results in significant price increases, continued Government support for renewables would no doubt be much more feasible. However, an important factor will be the 'politics of competition' versus the 'politics of the environment'. Having been driven this far in the utility industries, will competition succeed in driving out all subsidies, levies and NFFOs, thus doing nothing for the environment? Or will the 'politics of the environment' begin to take over and insist that more is done by way of energy efficiency and renewables?

Nevertheless, the final conclusion is that Government intentions post-NFFO5 are still unclear and, under the competitive privatized electricity system in England and Wales, many renewable energy technologies will have a problem surviving without any support. An obligation of PES to buy the electricity at a reasonable cost is likely to be a necessary requirement for renewables to continue their development.

9

The Winners and Losers So Far

Gordon MacKerron and Jim Watson

INTRODUCTION

Setting out the factual record of the winners and losers of the first six years of privatization is fairly straightforward. What is more difficult is to interpret the record. There are a number of reasons for this. First, while five full years of financial data are now available, it will take a further three years before the intended re-structuring will be complete in England and Wales. Furthermore, the regulator has only quite recently completed his first round of re-setting price controls in transmission, supply and distribution, and there has not yet been time to observe the full impact of these changes.

Second, there are inevitably problems of assessing the results that have occurred against what might have happened if the old public sector regime had continued. This is a familiar methodological problem in the social sciences but a particularly serious one in a situation where so many elements of the old system were simultaneously changed at privatization. An example is the problem of knowing how quickly the UK coal industry would have been run down if both electricity supply and coal industries had remained in public ownership. Third – and another result of the simultaneous changes in ownership, structure and regulation – is that even where it is possible to determine outcomes, it is difficult to know how to interpret their causes. If, for instance, industry costs have fallen, how far might this be attributable to ownership changes, how far to increased competition, how far to liberalization of fuel and equipment purchases, and how far to responses to new forms of regulatory control?

This chapter concentrates mostly on the two most obvious 'stakeholder' groups in the industry – consumers and shareholders. It then moves to a consideration of the third main direct stakeholder group, employees, and then in turn considers (briefly) environmental issues, the financial position of the Government, and the fate of the main technological and fuel interests in the industry: coal, nuclear power, renewables, gas and, finally, research and development

THE INTENDED WINNERS AND LOSERS

In the overt political agenda, as most clearly shown in the White Paper which announced electricity privatization, the main beneficiaries were to be electricity consumers. Competition would 'create downward pressures on costs and prices, and ensure that the customer... comes first'.[1] The vehicle for ensuring that customers would get these benefits was to be the area boards or distribution companies. The CEGB had largely frustrated the intention of the 1983 Energy Act that private producers should enter the industry (see Chapter 2), and the large public sector monopoly of generation was to be dismantled. The RECs on the other hand were smaller and closer to customers, and future investment in generation was to be 'driven by the distribution companies'[2] rather than by the generators. In the initial political settlement of the ESI at privatization, the price controls applied by the Government to RECs were certainly far from demanding, especially in distribution, where RECs were allowed to raise real price levels annually for the first five years (see Chapter 5).

Within the ESI, nuclear power, and the nuclear industry more widely, were clearly also intended to do well. A substantial section of the White Paper explained the benefits of nuclear power to consumers, and it talked of the 'nuclear programme' as something that should not be interrupted. To this end, a powerful system of protective controls were put in place to ensure that nuclear electricity would be bought and (initially) to ensure that new nuclear capacity could be built. The obligation on RECs to buy specified quantities of non-fossil electricity (in practice almost entirely nuclear) was written into the initial privatization proposals.

The White Paper also explicitly talked of privatization as a process that would produce an industry 'more responsive to the needs of ...

employees'.[3] However, employees represent a cost to employers, and costs were to be cut in the new ESI; given, too, the record of labour-shedding and limited growth in real wage levels in other privatized industries (especially British Telecom), it never seemed very likely that this rhetoric would be fulfilled in a way that employees would themselves have wished. While the Government perhaps necessarily spoke only of winners in the White Paper, it was unlikely that there would be no losers at all.

There was an implicit, partly hidden agenda in the electricity privatization (see Chapter 3) as well as an overt and manifest set of objectives. In common with all the main British privatizations, the electricity privatization was meant to encourage wider private shareholdings and the Government hoped that the new shareholders would suffer few financial surprises at least in the early years. Interestingly, shareholders were not even mentioned in the White Paper (except in a limited reference to the chances for employees to own shares in their own company). Within the implicit agenda, the coal mining industry (and the National Union of Mineworkers in particular) were not intended to do well, although as Chapter 6 makes clear, the continuing ownership of coal by the Government constrained the extent to which the Government allowed market forces full rein in the process of negotiating coal contracts with the new private generators.

Consumers – Early Winners?

A central issue in evaluating the results of privatization is the impact on consumers. There are two distinct questions here: standards of service, and prices. These are examined in turn, and it is also necessary to consider consumers in a number of groups. At least four groups need to be treated separately: domestic and small commercial consumers (below 100 kW) who remain captive to RECs until at least 1998; and small, medium and large industrial consumers, whose experiences have diverged from each other as well as from domestic tariff consumers.

Standards of Service

Consumer service can be measured along a number of different dimensions. While most regulatory attention has been given to domestic consumers, there are also important issues about supply reliability and frequency of supply interruption that concern all consumers. Most dimensions of consumer-service levels are an issue for the RECs in their role as distributors, rather than for RECs in their supply businesses for other competing suppliers. This is because: first, the physical availability of electricity is always under the control of the monopoly distribution business of the local REC; and second, in the important area of domestic consumers, all supply or retail activities remain the responsibility of RECs until at least 1998. There are, however, some issues of service levels for larger customers in the competitive market, where competing suppliers may introduce new or improved services of a financial nature. Standards of service were in general high before privatization (with a few exceptions, notably the disconnection policy): the main fear was therefore that standards might fall.

While there has been no direct competition in the below-100 kW market, OFFER publishes detailed annual figures comparing the performance of the different RECs in a wide range of service areas.[4] This falls well short of 'yardstick competition' (in which financial sticks and carrots are applied by regulators according to the comparative performance of different companies along various dimensions) but it provides some incentives for RECs to maintain and improve their performance in the service area. The existence of Electricity Consumers' Committees covering each REC franchise area may have some similar effects. The various aspects of service standards are now considered.

Has the New System Kept the Lights on?

The answer has been positive. At privatization there was a surplus of generating plant, and so it was unlikely that the new system would fail on account of plant shortages, at least in the very early years. The large volume of new investment in CCGTs made plant shortage seem even more remote: however, the new capacity was more than matched by closures of old plant, and surplus plant in early 1996 had reduced to 18 per cent over maximum demand, from 27 per cent in 1990.[5] There were times in the winter of 1995/96 when various factors including cold weather and higher than expected breakdowns of nuclear plants led to plant availability running only marginally above peak demands (see Chapter 4 on the Pool price 'spikes' in December 1995, indicating serious shortages of plant availability). Although plant shortage has not yet led to any power cuts, it is clear that the new system does not have mechanisms that ensure adequate plant supply over the longer term (see again Chapter 4, on the inadequacy of Pool incentive mechanisms as a way of ensuring that adequate generating capacity is available).

The other main potential cause of the lights going off is a breakdown in transmission or distribution systems. Here it was not initially clear how well the new system would work in terms of reliable delivery to consumers. However, performance in this area has been good: the high voltage system (the responsibility of the National Grid Company) has been reliable, and the RECs have generally improved their performance in terms of the number and duration of interruptions to supply. Between 1990/91 and 1993/94, the number of annual interruptions per 100 customers in England, Wales and Scotland fell from just over 120 to just under 100; over the same period and for the same area, the average number of minutes of supply lost per customer also fell, in this case from around 230 minutes to just over 100 minutes.[6] In other words, the number of interruptions to supply fell by just over 20 per cent, and the total duration of interruptions was slightly better than halved.

The overall record, then, is of some real improvement in the reliability of electricity supplies since 1990/91. It is important to compare this performance with system operation before privatization. In the case of the number of interruptions, the 1993/94 performance was better than the average performance over the previous ten years for 10 out of the 14 companies. In the case of the number of minutes lost per customer, the 1993/94 record shows improvement on the ten year average for 12 of the 14 companies.[7] The physical reliability of electricity – already very high – has therefore improved a little further over the past decade, with some continuing improvements in the post-privatization period.

Performance Standards

These can apply to all customers, but the main focus here is on the domestic market. OFFER has been active in this area, and has divided performance issues into Guaranteed Standards, where failure by a REC results in a penalty payment to consumers, and Overall Standards, where compensatory payments are not appropriate.[8] Such Standards were first introduced in 1991, tightened in 1993, in 1994 (to a limited extent), and again in 1995. The original levels of service required by the Standards reflected the levels of service performance being delivered by the nationalized industry just before privatization, and current Standards are therefore somewhat higher than the levels of public-sector achievement.

There are ten Guaranteed Standards, covering such issues as: restoring supplies following a fault; providing a supply and a meter; giving notice of interruptions; charge enquiries; and keeping appointments. In each case the RECs are required to respond within certain minimum periods, or pay compensation. Overall, RECs have consistently improved their performance in meeting Guaranteed Standards: the number of payments by RECs for failure to meet Guaranteed Standards fell by over a half, from 48 per 100,000 customers to 21 per 100,000 customers in the two years to 1994/95, despite a significant tightening in standards half-way through the period. Performance is by no means uniform across all the RECs – for example the East Midlands REC makes over 30 times as many payments per customer as London – but there is a widespread trend towards improvement. Nine of the twelve RECs reduced their payment levels between 1992/93 and 1994/95, and only Eastern and Yorkshire experienced significant rises.

There are ten Overall Standards, and here the requirement is expressed in the form that RECs should carry out particular tasks within a prescribed period in a specified proportion of all cases. The proportions range between 75 and 100 per cent, and vary both by task and by REC. Among the Overall Standards are: correcting voltage faults; connecting and reconnecting supplies; changing and reading meters; and dealing with customer correspondence. Standards have also been tightening here, but performance improvement has been marginally less uniform than for Guaranteed Standards. However, failure to meet a Standard is rare, and in 1994/95, only Midlands (5 cases), East Midlands (2), Manweb (1) and Eastern (1) actually failed to meet a particular Standard.

The general picture in the case of Standards of Performance is clearly one of further improvement in what was already a high level of customer service at privatization.

Codes of Practice

These are, broadly, the rules specified by OFFER on billing practices for domestic consumers; on services for the elderly and disabled; on advice relating to energy efficiency; and complaint procedures. In addition, disconnection policies are also covered in this section. Generally the Codes of Practice have been tightened over the five years to 1994/95, and the level of complaints received by OFFER against the companies has fallen from 16,677 in 1992 to 10,007 in 1994.[9] OFFER has pursued a vigorous policy of discouraging the companies from disconnecting consumers for non-payment of bills (while insisting that appropriate arrangements are made to pay arrears) and this policy has now been fully implemented. In 1989/90, 80,000 domestic consumers were disconnected, and the level of disconnection has been falling steadily and rapidly ever since, reaching the very low level of 1228 in 1994.[10] The policy of keeping consumers connected in no way guarantees that consumers can acquire even minimal levels of electricity, because pre-payment meters are normally installed in cases of non-payment. However, the new approach represents a more humane practice than the old policy of disconnection. Campaign groups had lobbied for an end to disconnections when the industry was in public ownership, but with little success: it is ironic that it took a regulated private industry to achieve a social goal that a public industry might have been expected to achieve easily.

Energy Efficiency

OFFER has a statutory responsibility to promote energy efficiency, and this is consequently an area where consumers might have expected substantial new activity for their benefit. In practice neither the companies nor OFFER have shown much interest in energy efficiency, especially because the initial forms of price control gave clear incentives to companies to maximize the volume of sales. There have been changes since the supply and distribution price reviews in 1994 and 1995 respectively. First, the formulae used to calculate REC revenues no longer give such a strong impetus (the 'volume incentive') to maximize sales, although on its own this is likely to be at best a weak incentive. Second, the supply price review included a new provision that a levy of £1 per customer a year for four years to 1998 (which will raise, including provision in Scotland, a total of nearly £100 m over the period) will be used on specific energy-saving projects.[11] The projects will be managed by the RECs, and OFFER has developed standards of performance for the companies to follow, having taken advice from the Energy Saving Trust, an independent non-statutory body given responsibility for overseeing UK energy savings, initially with special reference to carbon dioxide emission reductions.[12] Performance standards here include target savings in electricity for each REC over the four years, and a set of criteria to be applied to proposed projects. The criteria concern cost-effectiveness, a presumption in favour of consumers who are elderly or disabled, and an attempt to prioritize projects which put downward pressure on general charges. The thinking behind this latter criterion is that all consumers will contribute to the projects and that, to avoid discrimination, a high proportion of customers should see benefits in return. Early indications are that the standards will secure the objectives of the programme, but as OFFER itself argues, it is a limited programme and will at best make a marginal impact on energy use (the target savings amount in total to around 2 per cent of annual British consumption of electricity). In the area of energy efficiency, a Government lukewarm in its approach to electricity – combined with a slow and cautious regulatory approach – have given few tangible results to consumers as yet.

Service Levels to Consumers in the Competitive (above 100 kW) Market

The area for service competition is here quite small, because the physical availability of electricity depends on the distribution system, which remains a monopoly, REC-controlled business. Nevertheless there is evidence that competing suppliers have used improved financial services (eg making the billing period coincide with the consumer's own accounting cycle, and billing across all sites owned by the same company) as a weapon of competition. In addition metering is now a competitive business, and some improved services, as well as lower costs, are now also available for some customers. Nevertheless most large consumers still report that price is the most important element in the competitive market.

For consumers therefore there have been no alarms or surprises in the wide-ranging area of service levels. Standards of service were already high in the old state-owned system, with a limited number of exceptions such as the practice of disconnection for non-payment. In almost all areas of service provision there have been further improvements, though energy efficiency has been neglected until recently. Changes to service standards have mostly not been very noticeable in the aggregate, precisely because the earlier levels of service were already high. As for the future, the system installed by OFFER is likely to produce further steady, if unspectacular, improvements in standards of consumer service.

PRICES

While consumers are unlikely to notice service levels unless they slip notice-

Table 9.1 Average tariffs to standard domestic tariff consumers (England and Wales) (p/kWh)

	1989/90	*1990/91*	*1991/92*	*1992/93*	*1993/94*	*1994/95*	*1994/95*[1]
Average	7.95	8.8	8.93	8.60	8.60	8.53	9.21
Average (1989/90 price levels)	7.26	7.26	7.56	7.38	6.99	6.76	7.30
Highest tariff as percentage of lowest	112	114	114	113	119	119	119

Source: OFFER *Annual Report 1994* HC 432, 6 June 1995, figure 7, p 34.
1 Including 8% VAT.

ably, prices are generally of prime concern to all customers, and most customers are likely to judge the relative success of privatization by the impact it has had on the level of prices that they pay. Here it is necessary to divide consumers into four groups (tariff customers, especially households, and three different sizes of commercial and industrial consumers).

Politically, the most significant group has been the 22 million household consumers (in England and Wales) who have remained captive for retail supply to their host REC. The background to the evolution of tariffs (see Chapter 2) is the significant increase in tariffs in the two years between 1987 and 1989. The Government increased the target rate of return for the ESI from 2.75 per cent per year for the period 1985/86–1987/88 to 4.75 per cent for 1989/90 (the last full year of public ownership),[13] and this meant that domestic tariffs had to rise by 15 per cent (7 per cent in real, inflation-adjusted terms) in the two years to 1989. Consumers had therefore experienced a noticeable rise in real tariffs, unrelated to underlying cost levels, in the run-up to privatization. This is the background against which to view Table 9.1, which shows the evolution of the average, standard-rate domestic tariff between 1989/90 and 1994/95 for the English and Welsh companies.

In inflation-adjusted terms and ignoring VAT for the moment, the progress of domestic tariffs has been of a small (4 per cent) rise in

1991/92, followed by an accelerating decline, by a total of almost 11 per cent between 1991/92 and 1994/95. The most pronounced annual fall in tariffs (just over 5 per cent) occurred in 1993/94, and reflected the new April 1993 coal contracts (see Chapter 6) which reduced the price of UK coal and reduced the contracted volumes. Without the introduction of VAT on electricity, domestic tariffs in 1994/95 would have stood some 7 per cent lower than they had in 1991/92. However the introduction of 8 per cent VAT on domestic electricity bills in April 1994 meant that actual prices were, by 1994/95, at almost exactly 1989/90 levels. Domestic consumers had neither very obviously gained nor lost over this five-year period in absolute terms.

Prospects beyond the end of 1994/95 are for significant further falls in domestic price levels in England and Wales. The distribution price review imposed a one-off cut in distribution charges of 11–17 per cent in April 1995, and a further distribution price cut in the range 10–13 per cent will be applied in April 1996 (see Chapter 5). As distribution accounts for roughly a quarter of the domestic retail price of electricity, the combined effect of these two cuts will be to reduce domestic bills in the range of 6–7 per cent over the two years concerned. A second price cut should follow the Government's intention to remove the nuclear component of the fossil fuel levy in England and Wales some time in late 1996. Originally the Government announced that this process would produce a fall in prices of 8 or 9 per cent (implying that the levy would fall from the current rate of 10 per cent to 1–2 per cent). However it has been announced subsequently that arrears of fossil fuel levy proceeds will need to be collected and that the full reduction will not take place as early as first planned (see Chapter 7). The continued collection of the nuclear part of the fossil fuel levy is in practice a device to help the Treasury rather than Nuclear Electric (its immediate beneficiary). This is because Nuclear Electric has no further need of the levy and simply passes it through to its Government-held bank account or its holdings of Government stock, thus reducing the public sector borrowing requirement. It now seems likely that the rate will be cut to 3.5–4.5 per cent in late 1996, leading directly to a 5.5 to 6.5 per cent reduction in bills.

The combined effect of these changes will be that domestic consumer bills will fall by some 12–13 per cent in the two years from April 1995. In addition, in early 1996 all consumers in England and Wales are being paid a flat, one-off £50 rebate as a share in the profits that RECs have made from the flotation of the National Grid Company in December 1995. This has resulted from political pressure on the RECs (the previous joint owners of National Grid) to share some of their large flotation windfall directly with consumers. The implied market value of National Grid on privatization as a wholly owned subsidiary of the RECs in 1990 was around £1 bn, and its opening value as an independent company in December 1995 (including its now separated pumped storage stations) was around £4.2 bn.[14] The payment of £50 per customer represents a cost to the RECs of £1.1 bn, leaving a gain to the RECs on the flotation of National Grid that still exceeds £2 bn.

Domestic consumer prices in England and Wales should therefore fall substantially from 1996 to 2000. A share of the economic surplus that in the first five years of privatization was almost entirely captured by the new owners of the ESI (see below) is therefore starting to go to consumers in the second half of the 1990s.

Another potentially significant price issue for the domestic market is the variability in tariff by location (and in all probability after 1998, by individual customer). At present, and following earlier practice under public ownership, all consumers in a particular tariff category have been charged the same price within each REC, with tariffs varying between RECs for similar customers. Table 9.1 shows some widening in the range of prices charged. In 1989/90 the highest domestic tariff ((Manweb's at 7.66p/kWh) was 12 per cent higher than the lowest (Eastern at 6.84p/kWh). The differentials showed little change until the last two financial years and by 1994/95 SWALEC's tariff (10.21p/kWh) was 19 per cent above Norweb's (8.53p/kWh).[15] An increasingly competitive system – RECs now negotiate freely and individually for wholesale contracts – will tend to produce more variation in consumer pricing. This tendency towards wider differentials in consumer charges already apparent will almost certainly become more pronounced if, as planned for April 1998, all domestic consumers gain free choice of supplier.

It is more difficult to give a comprehensive review of prices charged to industrial consumers. This is because the competitive market, since the limit was reduced from 1 MW to 100 kW in April 1994, has represented half of all electricity sales (and the great majority of all commercial and industrial sales) and in this market most transactions are commercially confidential. Some aggregate data[16] are available (Table 9.2) for three size categories of industrial consumer. These categories roughly correspond to the below-100 kW, 100kW-to-1 MW and above-1 MW market segments.[17] Problems with Table 9.2 are that there is no way to assess variation and

Table 9.2 Average electricity prices for industrial consumers (1988 prices) (p/kWh)

	1989	*1990*	*1991*	*1992*	*1993*	*1994*	*1995*
Small consumers (<880 MWh per year)	5.13	5.38	5.55	5.58	5.26	4.96	4.72
Medium Consumers (880–8800 MWh per year)	4.19	3.89	3.72	3.72	3.75	3.53	3.27
Large consumers (>8800 MWh per year)	3.09	2.86	2.74	2.82	2.91	2.80	2.58

Source: Department of Trade and Industry *Energy Trends* table 28, various issues, including January 1996.

ranges within each category, and that the very largest consumers (who have had a different experience of prices) cannot be distinguished from other large customers.

The large-consumer market became competitive in April 1990, and Table 9.2 shows a drop of 7 per cent in the real price paid between 1989 and 1990, mostly as a consequence of this introduction of competition. The level of price reduction over the whole period 1989 to 1995 is 17 per cent. This does not in practice reflect a uniform experience in this segment of the market. The very largest consumers benefited from politically negotiated schemes before privatization (QUICS) which gave them below-cost electricity (by the device of supposing that the UK coal which supplied the power they used was priced at world levels, rather than at the much higher prices which reflected UK coal costs). This scheme was not withdrawn fully until 1991, but the effect of its withdrawal was that some very large consumers experienced sharp price *rises* in 1991. Thus the 17 per cent reduction in prices across the whole sector is the product of greater price falls for the smaller consumers in this category (typically perhaps by 20 per cent) combined with limited price falls for the very largest users.

The medium-consumer market became competitive in 1994, and the main price reductions came in the two years following the opening of competition – of about 13 per cent in real terms. Between 1989 and 1994, prices in this segment had declined by an average of only about 2 per cent annually. On average, then, medium-sized consumers have seen significant overall price reductions of almost 22 per cent over the six-year period to 1995, with the main impact since the introduction of competition in 1994.

Smaller industrial consumers have generally remained on tariff terms and have been supplied throughout by their local REC. Their experience – apart from the absence of VAT in this market – has been very similar to that of domestic tariff customers. Prices rose significantly in the early years of privatization – by almost 9 per cent between 1989 and 1992, and then fell by 15 per cent between 1992 and 1995. At 1995 therefore, prices to smaller industrial and commercial consumers stood around 8 per cent below the 1988 level. As in the case of domestic consumers, therefore, the smaller industrial sector was only beginning to see price reductions after several years of privatization.

The overall picture for consumers on prices is that domestic and small industrial consumers initially experienced modest price rises followed by slightly larger price reductions (an effect masked in the domestic market by the imposition of VAT at 8 per cent in April 1994). Medium-to-large industrial consumers have fared better in price terms and most have seen noticeable price reductions following the successive introduction of competition in the 1 MW and 100 kW markets in 1990 and 1994. The very largest consumers, however, have barely recovered from the removal, in 1991, of the subsidy scheme which previously gave them below-cost electricity.

Shareholders – the Real Winners?

It has been widely observed that the electricity companies have become substantially more profitable since privatization, and that shareholders have acquired the lion's share of the economic rent available at the expense of consumers and taxpayers. This section tests the extent to which the record bears out this popular view. The first indicator, simple but useful, is the increase in the value of shares in the companies since flotation. For the RECs, floated in December 1990, the increase in share values from that time to August 1995 has increased the value of the companies by almost exactly 200 per cent – from £5.18 bn to £15.48 bn. In the case of the two large private generators, the increase in share values from flotation in March 1991 to August 1995 was slightly less – an increase of 171 per cent over a slightly shorter period, from £3.60 bn to £9.79 bn.[18] This is a rate of growth of share value, for both groups of company, of almost exactly 25 per cent a year for a period of four to five years, at a time of generally restricted growth in market values. In the four years between January 1991 and December 1994, the value of shares in the electricity sector exceeded the average (represented by the Financial Times All Share index) by 80 per cent, or 16 per cent a year above average growth.[19] This is clearly an exceptionally profitable performance.

How much have profits risen? The choice of different conventions and assumptions can give significantly different results, and the main considerations and choices here have been as follows. 1990/91 has been taken as the base year, partly because this was the first financial year following the vesting of the companies in April 1990, and partly because the 1989/90 figures are distorted, for the generators, by one-off exceptional payments. Where possible, the pre-tax profit level has been used because this gives the fairest general view of the financial health of the companies. However, for the RECs it has been necessary to use the (normally higher) measure of operating profit. This is because to obtain a breakdown between profits in the monopoly distribution business and the increasingly competitive supply business it is necessary to refer to the regulatory accounts,[20] which are only expressed in operating profit terms. For Nuclear Electric it is also more appropriate to use operating profit: while pre-tax profits are available they are distorted by many random fluctuations from year to year.[21] The figures have been divided into three categories: the competitive market, which includes National Power, PowerGen and the supply businesses of the RECs; the monopoly market, which includes the distribution businesses of the RECs and the National Grid Company; and Nuclear Electric, which, with its over £1 bn annual consumer subsidy does not fit neatly into normal commercial categories (Table 9.3).

The figures in Table 9.3 suggest a number of conclusions. First, profits in the competitive part of the industry rose, over the four-year period, by a total of 75 per cent, or a compound rate of 15 per cent annually. This was topped, marginally, by the monopolistic part of the industry, where profits rose by 82 per cent, or 16 per cent annually. This very strong profit performance in the monopoly part of the industry has been of particular public concern, as profit levels are related quite closely to price regulation, which has been widely seen as very lenient. National Grid, with an assured income stream and very low risks indeed, was still earning over 9 per cent annual rate of return on current cost accounts in 1994/95[22] – the second full year *after* its price cap was re-set in April 1993.

Table 9.3 Profitability of the electricity companies, 1990/91–1994/95 (£m)

	1990/91	*1992/93*	*1994/95*
The competitive market			
National Power	479	580	705
PowerGen	272	425	545
RECs' supply businesses	100	173	240
Total	*851*	*1178*	*1490*
The monopoly market			
National Grid Company	386	533	611
RECs' distribution businesses	914	1501	1753
Total	*1300*	*2034*	*2364*
Grand total (private companies)	*2151*	*3212*	*3854*
Nuclear Electric	*326*	*664*	*1218*

Sources: *Annual Reports and Accounts* of the various companies
Figures for National Power, PowerGen and National Grid are pre-tax profits. For the RECs and Nuclear Electric, they are operating profit.

Second, profit growth in both competitive and monopolistic parts of the industry was most rapid in the first two years, and has clearly slowed down in the two years to 1994/95. Table 9.3 shows that the generators raised profits by 38 per cent in the first two years and 26 per cent in the two most recent years. National Grid and the distribution businesses saw an even more rapid reduction in profit growth – from 56 per cent in the first two years to only 16 per cent in the two years to 1994/95.

Third, Nuclear Electric was able, by 1994/95 to make sufficient profit as no longer to need the average £1.2 bn fossil fuel levy payments that it has received annually. Nevertheless consumers still pay the levy at 10 per cent, and the levy now represents a significant re-distribution away from consumers and towards the owners of Nuclear Electric (the Government – see Chapter 7).

Fourth, the figures for total profits also represent very large increases in the rates of return on assets, as the asset values of the industry did not change markedly during the four-year period covered.

Finally, the total profitability of the private and public industry in

England and Wales has roughly doubled from just under £2.5 bn to just over £5 bn in the four years to 1994/95.

With share values and profit performance of the kind just outlined in a fairly flat market, and with significant monopoly powers remaining even in the competitive part of the industry, the perception that the privatization has been too generous to shareholders is very widespread. It led to the political pressure on RECs to pay the £50 rebate to consumers after flotation of the National Grid Company, even though this still left the RECs with a windfall of over £2 bn.

It is therefore clear that as between consumers and shareholders, the first years of privatization have benefited shareholders to a remarkable degree. Most consumers have seen only very marginal improvements in price, together with small (but welcome) improvements in already high standards of service. Shareholders, by contrast, have made extremely large gains.

If profits rise very steeply while prices dip slightly, costs must clearly have fallen very substantially. If these cost reductions could be mainly attributed to efficiency gains consequent on the new privatized incentives to cost-cutting, there could be a case for supposing that shareholders might legitimately be the principal beneficiaries. In practice, much the greater part of the large cost reductions have had their origin outside the electricity privatization – mainly (see Chapter 6) sharp falls in the price of coal. Precise quantification of this effect is impossible given the data available, but broad indications can be inferred.

In the first years of electricity privatization, coal prices to the generators fell comparatively slowly. Chapter 6 shows that the decline in the level of real (inflation-adjusted) coal prices to generators was roughly 4 per cent annually from 1989/90 to 1992/93. These changes resulted from the three-year coal contracts which the Government enforced between the generators and BCC as part of the privatization process. These contracts ran from April 1990 to March 1993 and specified minimum quantities of coal to be supplied, with prices even in 1992/93 that were at levels well above the international level (see Chapter 6). The burden of these relatively high-priced coal contracts was entirely borne by domestic consumers. This was because the excess costs of BCC coal were directly passed on to the RECs in so-called back-to-back contracts for electricity supply, and the RECs in turn passed these costs on to their captive domestic consumers.

The new coal contracts from April 1993 signalled a real change and enforced large cuts in both prices and volumes of coal contracted to the generators. The Government was less able to be influential in this second round of coal negotiations because the generators were now less subject to Government control than in the pre-privatization negotiations. After much argument and a political storm over proposed pit closures, the new contracts specified reductions in volumes from 65 million tonnes in 1992/93 to 40 million tonnes in 1993/94, and 30 million tonnes every year thereafter until 1997/98.[23] Prices were also to fall substantially in 1993/94 (by about £8/tonne, or 18 per cent in one year) and would continue to fall by a further 12 per cent over the period to 1998.

The financial impact in 1993/94 on the two generators was even greater than this new contract alone implied. National Power and PowerGen had a huge coal stock 'overhang' by the end of 1992[24] because the first three-year contract had obliged them to buy more BCC coal than they could burn (see Chapter 6). In addition, their annual coal needs were declining as they began to lose market share. This meant that the extent of their need to replace the 25 million tonnes of coal no longer purchased from BCC was very limited. Total UK steam coal imports in 1993/94 amounted to only just over 6 million tonnes,[25] and for the purpose of this calculation it is assumed that the English and Welsh generators may have bought as much as 5 million tonnes, and that they might have paid £34.50/tonne for these imports.[26]

The balance of coal costs to the generators as between 1992/93 and 1993/94 was roughly as follows. In 1992/93, 65 million tonnes of coal at £42.50/tonne from BCC (see Chapter 6) would have cost £2763 m. In 1993/94, 40 million tonnes at £34.50/tonne would have cost £1380 m. In addition, 5 million tonnes of imports at £34.50/tonne would have cost £173 m, making a total cost in 1993/94 of £1553 m. This represents a saving in coal costs of a little over £1209 m. It is true that in 1993/94 the generators were increasing their gas purchases, but this would only slightly have offset the savings on coal.

This calculation is necessarily indicative only, but the order of magnitude is confirmed by the observation that the combined operating costs of National Power and PowerGen fell by £1126 m in 1993/94 over 1992/93 (Table 9.4).

Table 9.4 also shows that 1992/93 was the only other year in which the combined operating costs of the two companies fell (by £356 m). This was another year of reductions in the value of coal purchases – a fall in UK contracted volume by 5 million tonnes, saving over £200 m. In the other three years since 1989/90, the (non-inflation adjusted) operating costs of

Table 9.4 Operating costs of National Power and PowerGen, 1989/90 – 1994/95 (£m)

	1989/90	*1990/91*	*1991/92*	*1992/93*	*1993/94*	*1994/95*
National Power	3542	3951	4176	3749	2907	3184
Power-Gen	2216	2384	2677	2739	2455	2354
Total	*5758*	*6335*	*6853*	*6488*	*5362*	*5538*

Source: Company Annual Reports and Accounts.

the two generators have collectively risen rather than fallen.

It is of course arguable that the rate of decline in coal prices and volumes might have been lower (especially after 1993) if the ESI had remained in public ownership, and that to this extent the falling coal bill is in part attributable to privatization of the ESI. It is highly likely that in the absence of ESI privatization more coal would have been needed from 1993 onwards (given the probable absence under public ownership of a major 'dash for gas') but BCC had already embarked in the late 1980s on a radical programme of cutting capacity, and cost and price reduction. While some part of the fall in the coal bill may be attributable to ESI privatization, none of it may reasonably be held to be the result of improvements in internal efficiency within the electricity business.

Non-fuel costs in fact have declined rather modestly in the last five years, as Table 9.4 implies. National Power have reported a fall in non-fuel costs of 10 per cent in the five years of their existence[27] (a similar rate to that of PowerGen). This annual rate of 2 per cent reduction is comparable to what the CEGB was achieving in its latter years (see Chapter 2), and can only explain a very minor part of the total fall in real cost levels.

EMPLOYMENT

Among the non-fuel costs, it is clearly in employment terms that the new companies have made their major efforts to reduce costs. It is certainly true that the 'head-count' has fallen rapidly among the generators (including Nuclear Electric) and more gradually in the RECs (Table 9.5). The generators have certainly reduced their employment levels rapidly.

Table 9.5 Employees in the privatized ESI (average numbers in each financial year)

	1989/90	*1990/91*	*1991/92*	*1992/93*	*1993/94*	*1994/95*
A Generators						
National Power	16,977	15,713	13,277	9934	6955	5447
Power-Gen	9430	8840	7771	5715	4782	4171
Nuclear Electric	14,415	13,924	13,300	12,283	10,728	9426
Total	*40,822*	*38,477*	*34,348*	*27,932*	*22,465*	*19,044*
B The network and supply						
National Grid Company	6442	6550	6217	5666	5127	4871
The RECs	82,485	82,288	81,135	77,329	71,473	65,787
Total	*89,127*	*88,838*	*87,352*	*82,995*	*76,600*	*70,658*
Grand Total	*129,949*	*127,315*	*121,700*	*110,887*	*99,065*	*89,702*

Sources: Company Annual Reports and Accounts

National Power, for instance, employed, in 1994/95, fewer than a third of the people that they did in 1989/90 and, even in the state sector, Nuclear Electric shed 5000 jobs up to 1994/95. However, this is not such a remarkable change as is often claimed. First, the private generators have closed a large amount of old and relatively labour-intensive plant. Second, many functions previously performed in-house are now subcontracted to outside firms (eg plant maintenance). This means that the quantity of labour applied to the generators' operations has fallen significantly less quickly than their head-counts. (However, it is important to remember that the terms and conditions applying to those now contracted-in to undertake maintenance and other work will in most cases be substantially inferior to those applying to the staff of the electricity companies.) Third, the figures presented in Table 9.5 exclude employment in the new IPPs. While CCGTs require modest amounts of labour compared to coal-fired stations, the existence of a significant independent sector means that generation employment is, by 1994/95, at least a little higher than appears from Table 9.5.

In any case, labour is an almost insignificant element in generators' costs (for instance for PowerGen, less than 6 per cent of operating costs in 1994). Consequently, large increases in labour productivity, such as the 119 per cent improvement in the case of National Power in the five years to 1994/95,[28] translate into a saving of only 10 per cent of all non-fuel costs. While the generators have expended much effort on reducing their head-counts, the impact on costs and profitability has been quite limited compared with the impact of lower fuel input costs.

In the case of the RECs and National Grid, job losses have been much more modest, though they have accelerated markedly in the case of the RECs in the two financial years to 1994/95, partly in response to the new price controls between 1993 and 1995. While labour is a more important element in REC costs than for the generators (averaging 10 to 13 per cent across the REC spectrum),[29] it is still a relatively minor element in total costs. In any case, the job losses in the monopoly part of the industry are probably not much higher than the run of losses in other low-growth UK businesses over the same period. In these network activities employment fell by just over 21 per cent in five years, an annual rate of almost exactly 4 per cent.

Finally, compulsory redundancy has been rare in the process of reducing labour levels, and a high proportion of employees leaving have received generous severance terms. While apparent job losses of about 31 per cent in the aggregate are considerable, it seems that the scale and significance of job losses in the ESI has sometimes been over-played. While employees of the ESI have not fared particularly well, their experience has not been – by present British standards – very exceptional.

It is much more difficult to over-play the extent of the gains made by the most senior members of the ESI. Executives in virtually all the electricity companies have experienced very large increases in their pay, and in many cases have all been furnished with generous share options. Annual reports of the RECs show that the typical level of annual directors' salary rose from around £60,000 in 1989/90 to £210,000– £230,000 in 1994/95. In 1988 the Chairman of the CEGB received around £100,000: the combined salaries of the Chief Executives of National Power and PowerGen had reached almost £850,000 by 1994/95.[30] Even more distressing to much public opinion has been the extent of gains through share options. The Chief Executives of the two generators and of the

National Grid Company are now reported to be millionaires as a result of these options,[31] and very large gains have been made by many ESI directors by this route. The Chairman of the National Grid Company made gains approaching £2 m from the flotation of his company in December 1995.[32] While there may seem justification for higher pay in areas where there is more competition and risk, it is difficult to see why the same directors, of the same companies in the monopoly activities of distribution and transmission, should raise their incomes by large multiples. Much of the adverse public reaction to the newly privatized system has focused on the very large increases to remuneration packages that the most senior people in the industry have mostly voted for themselves.

The Environment

The environmental performance of the privatized ESI has mostly been good. This has very little to do with explicit environmental policy and almost everything to do with the fortunate coincidence that the commercially most attractive fuel – natural gas – happened to possess substantial environmental advantages, along several dimensions, over the coal it mostly replaced. The two dominant environmental issues have been sulphur emission controls and carbon dioxide stabilization strategies.

In the case of sulphur, the 1988 Large Combustion Plant Directive (LCPD) of the EU requires the UK to cut sulphur emissions by 60 per cent over their 1980 levels by 2003 (Chapter 5). This in turn has been largely superseded by the UN Economic Commission for Europe's Sulphur Protocol which for the UK specifies a 70 per cent cut by 2005, followed by 80 per cent by 2010. When the LCPD was concluded it appeared that flue gas desulphurization technology would need to be retrofitted, at substantial expense, to as much as 12 GW of English and Welsh electricity capacity.

As soon as privatization became certain, after the passage of the 1989 Electricity Act, it quickly became clear not only that all existing and potential generators would no longer invest in coal, but that a significant new level of gas-fired capacity would come into existence, virtually all of it replacing coal use that would otherwise have continued for a long period into the future (see Chapter 4). As gas is essentially free of sulphur, the impact of this changed investment strategy was dramatic. It became clear that no more flue gas desulphurization would be needed after the 6 GW already committed, and it also became evident by 1993 that the surge in new gas use for power generation would allow Britain to over-fulfil its target emission levels under the LCPD for 2003.

The rapid and large-scale switch to gas also proved of great benefit to Britain's other major international environmental commitment – stabilizing carbon dioxide emissions at 1990 levels in the year 2000 (see Chapter 5). When the Government was first committed to this target it envisaged the need to look for savings in emission levels across several energy and industrial sectors. A projected level of 168 million tonnes of year 2000 emissions would need to be cut by 10 million tonnes. The Energy Saving Trust was to manage the process, in household and industrial sectors, of achieving the required savings. Gas again came to the rescue: it leads to less than 60 per cent of the carbon emissions that an equivalent heat value of coal will give off, and in the combined-cycle mode, the higher efficiency of gas brings further gains. In practice then, CCGTs lead to less than half the carbon dioxide emissions than those that would arise if existing coal-fired stations produced the same quantity of electricity. By the late 1990s (see Chapter 6) CCGTs will have displaced more than 40 million tonnes of other fuel (nearly all of it coal). This change will, on its own and with much to spare, save the 10 million tonnes of carbon emissions that the Government needs to achieve. This seems to be the main cause of the very limited funding now available to the Energy Saving Trust compared to original expectations.

THE TREASURY

An important issue is the changes in financial flows between the electricity supply industry and Government as a result of privatization. As in other areas it is impossible to reconstruct what would have happened to these flows if public-sector ownership had continued, but some indications of actual experience before and after privatization can be given.

In the last normal year before privatization (1987/88) the ESI in England and Wales made a net contribution to the public finances of £1.47 bn, consisting of £356 m of Corporation Tax, £304 m of interest repayment, and £808 m of post-tax profit.[33] Through the 1980s the industry had been increasing the net sums it paid in to public finances, and there seem good chances that this trend might have continued in the absence of privatization.

The flotations of the privatized companies raised money directly for the Treasury over a number of years. The sale of the RECs raised £4.8 bn directly (plus staged payments of £2.8 bn of debt put in to the opening balance sheets to be repaid to Government in due course).[34] The value of National Power and PowerGen (sold in two stages) amounted to £6.3 bn, plus £868 m in debt.[35] The total sale value of the England and Wales industry – including debt repayments to be made over several years – was therefore approximately £14.8 bn, including £11.1 bn of immediate cash.[36]

This is not the end of the balance – the new companies have, as shown earlier, been highly profitable, and have therefore contributed quite large sums to the Treasury in taxes. Adding all the RECs to National Grid and the two private generators, total tax paid in 1990/91 amounted to £693 m, rising to £1014 m in 1994/95. In addition to this, the Treasury has been acquiring larger and larger sums through the share of Nuclear Electric's receipts of the fossil fuel levy that the company cannot use. By March 1995, the Government held some £2 bn in cash balances for Nuclear Electric, £1.5 bn of which had come directly from the levy.[37] The net balance of advantage to the Treasury from the privatization of electricity seems on the whole to have been rather favourable – there has not been an apparently large loss of revenue inflow (thanks to the profitability of the industry), and the sale proceeds were substantial.

THE FUELS AND THE TECHNOLOGIES

Privatization has clearly created winners and losers among the fuels and technologies – and their associated interests – that are used in electricity supply. In terms of new investment choices, the bare outline is stark. Before privatization, the only contenders for new investment were large coal and nuclear plants (see Chapter 2). After privatization, the only investments made were in gas-fired CCGTs, and (because of consumer subsidy) renewable energy. This remarkable turnabout can in large part be explained by changes in the discount rate applied to investment appraisals. In the public sector the CEGB had used a 5 per cent rate, following advice from the Treasury, until 1989, at which point the public sector rate was raised to 8 per cent (see Chapter 2). In the private sector, rates of 10 to 15 per cent are normal, and the effect of these very steep increases in the cost of capital is that the choice of investment is radically shifted. While low discount rates favour capital-intensive options (like nuclear power and to a lesser extent coal-firing) recent and much higher rates favour options – like combined-cycle gas – with low capital costs within the total generating cost structure. This change would undoubtedly have been a major disincentive to renewable energy – which is heavily capital-intensive – had not new policies favoured renewables development.

Nuclear power has done much less well than intended at the start of the process. The continued operation of existing reactors has been safeguarded by a strong regime of financial subsidy, but the process of privatization exposed the extent of the costs and risks of new nuclear investment in a commercial climate (Chapter 7). The original intention to use the NFFO to force RECs to buy power from new nuclear stations proved unworkable, and the White Paper on nuclear power of May 1995 simply confirmed what was already abundantly clear – that in a competitive private market, no one would invest in new nuclear capacity, and that, further, the Government had no intention of providing the state subsidy that would be needed to ensure new construction of nuclear plants (see Chapter 7).

Coal was always destined to do badly (Chapter 6). The only important question was how quickly the UK industry would be run down. In the 1990–1993 period the pace of this run-down was deliberately slowed in the contractual arrangements made between BCC and the generators; but this only led more or less directly to the crisis of late 1992, when BCC announced that 31 out of 50 deep mines would have to be closed within six months. While it could reasonably be anticipated at the time of privatization that new coal-based investment was unlikely, what could not be foreseen at the time of privatization was the speed and extent of the 'dash for gas', and this was an indirect rather than an inherent property of the post-privatization regime. While coal output levels for electricity generation will be held at an annual 30 million tonnes in the privatized coal industry at least until 1998, the damage done to employment in the coal industry far outweighs the job losses in the electricity supply industry. Rapidly increasing productivity and falling output levels have meant a fall in coal mining employment of enormous dimensions. This was a long-term process – employment in the nationalized coal industry fell from 294,000 in 1981/82 to only 85,000 in 1989/90. The decline continued precipitously after electricity privatization and the total number working at BCC fell to only 13,000 by the end of 1994.[38] Finally, of course, the British coal industry was privatized in early 1995.

It is unnecessary to say much here about the rise of gas. From a position in which gas use in power generation was negligible at privatization, there has been a phenomenal growth in CCGT construction and in gas use. At least 13 GW of CCGT capacity will be on stream by 1997 (Chapter 6), and by 1997/98, gas use in power generation will probably account for over 40 million tonnes of coal equivalent. Such a rapid transformation could not have been foreseen in 1989. It is of course partly a product of electricity privatization and the new climate of short-term returns on investment, but also importantly explained by the 'gas bubble' and the way in which the power station market provided an ideal vehicle for allowing new suppliers of gas to reduce the market dominance of British Gas.

Combined heat and power (CHP) schemes have done well in the privatized era. (In CHP the heat produced in electricity generation is used, usually for industrial purposes, rather than wasted, and overall thermal efficiency is normally twice as high as for conventional steam turbine systems of electricity production.) The Government has set a target of 5000 MW for the installation of CHP capacity by 2000, and between 1991 and 1994 installed capacity rose from 2312 MW to 3141 MW.[39] The growth in CHP has a number of origins. Cheap gas has been a prime stimulus, as has the improved underlying economics of small-scale gas-using systems, but some of the explanations are more directly connected to electricity privatization. Among these, the two most important are the improved (non-discriminatory) terms of sale of CHP electricity surpluses to the network, enforced by the new regulatory system, and the active marketing of CHP schemes by many of the privatized electricity companies.[40]

Renewables have in many ways done exceptionally well out of privatization, though largely fortuitously. Before 1989 no Government support had been given for commercial development of renewables (as opposed to some support for R&D), but this was rapidly reversed under the terms of the NFFO (Chapter 8). The result has been a rapid deployment of renewables at initially high, but subsequently declining, subsidy levels. Such support was initially born mostly out of a desire to lend some wider rationale to the subsidy that nuclear power required, and was not in any way an inherent part of the privatization. However, increasing environmental commitments after privatization (eg to carbon emission reductions) might well have led to a significant programme of renewables support irrespective of electricity-sector reform (Chapter 8). The acceleration of renewables development remains a major, if unexpected, outcome of the particular privatization process followed in Britain, though the future of renewables development remains uncertain if protection is withdrawn.

Finally, support for technology in general, in the form of research and development, has been a major casualty of the privatization process. Table 9.6 shows the evolution of expenditure on R&D over the privatization period. The aggregate picture is clearly one of a steep fall in R&D spend. However, the only clear trend beyond the aggregate is that a sharp fall in nuclear R&D (by £68 m from 1989/90 to 1994/95) accounts for slightly more than the whole of the total reduction in expenditure (which was down £60 m over the same period). This reduction in nuclear R&D is exactly what would be expected in an environment in which it became increasingly obvious that new nuclear investment was exceptionally unlikely. The bulk of the £50 m or so of current annual R&D by Nuclear Electric is essentially 'troubleshooting' work for operating reactors. No clear trend emerges for the two private generators or National Grid, and there is certainly no evidence of a decline. This is at odds with the observation that there has been a dismantling of much of the R&D infrastructure previously owned by generators and the national grid. Thus National Power closed its Leatherhead laboratory which had previously employed some 500 people, and kept only 50 or so of the former staff on its books. The answer seems to be that – as in other areas of employment – the industry has reduced its head-count much more radically than the amount of work done. In other words, a much higher proportion of R&D than before is now done for the industry by outside contractors, and this work is managed by much smaller numbers of internal R&D staff. The nature of much of the R&D undertaken has also switched to either short-term operational issues, or – to a significant extent – to research and development aimed at ensuring environmental compliance.[41]

CONCLUSIONS

Since privatization, costs in the electricity supply industry have fallen substantially, mainly due to reductions in the price of coal, with a much more minor contribution from a reduced labour force. This has given rise to large new economic rents. For reasons explored below, the new companies were able to avoid distributing almost all of these economic rents to consumers (certainly to domestic consumers) and the gains were therefore mostly captured by the new shareholders.

The political acceptability of this has been helped by the fact that while consumers have had few price benefits, they are in general no worse

Table 9.6 Research and development expenditure by the generators and the grid (£m)

	1988/89	*1989/90*	*1990/91*	*1991/92*	*1992/93*	*1993/94*	*1994/95*
CEGB	201	–	–	–	–	–	–
National Power	–	22	26	17	20	26	26
Power-Gen	–	5	14	12	10	9	8
Nuclear Electric	–	116	95	71	64	51	48
National Grid	–	7	7	8	8	8	8
Total	*201*	*150*	*143*	*108*	*102*	*94*	*90*

Sources: Company Annual Reports and Accounts.
Figures for RECs are not included because many do not report R&D spending. Those which do report expenditure in their reports and accounts generally show levels around £1 m annually (in no case falling short of £0.5 m or exceeding £1.9 m)

off in absolute terms than before privatization, and in some cases (eg middle-sized industrial consumers) noticeably better off. Service standards have generally risen a little from already high levels, and there are now clear signs that consumers will see some real gains in terms of falling prices in the next year or so. While employment in the new electricity companies has certainly fallen, this has been at rates that do not seem much out of line with the generality of experience in other low-growth industrial activities in the UK. The principal losers in employment terms have undoubtedly been coal miners.

More interesting – and difficult – is the question of what exactly has caused this very skewed distribution of benefits. The problem here is the simultaneity of so many changes in the England and Wales system – privatization, plus re-structuring and some competition, plus liberalization of fuel and equipment markets, plus new forms of regulation.

The most interesting question of all is: how have the companies been able to make such continuous increases in profit levels? This chapter suggests that there seem to be two broad answers, rather different in the competitive and monopolistic parts of the market. In the generation business, where competition was meant to be a main organizing principle of the new structure, costs have fallen significantly, almost entirely because of major reductions in coal prices. In a competitive market, these cost reductions would have been mostly passed on to distributors and thereafter consumers. In fact, generators' revenues have declined much less rapidly than their costs. In National Power's case, operating costs fell by a total of £904 m between 1990/91 and 1994/95, while revenues reduced by only £425 m.[42] This pattern strongly suggests that it was the substantial remaining monopoly powers of the two big generators that helped them retain a large share of their (mostly fortuitous) reductions in cost.

In the case of the RECs and NGC, the explanation is much more in the area of regulation. The monopoly areas of transmission and distribution have been price-capped throughout, and the main mechanism for raising profit levels here seems to have been the setting of very lenient price controls in the first round, so that RECs were in 11 out of 12 cases allowed to raise their real charges for five consecutive years (Chapter 5). In a situation where some cost-cutting was possible, this inevitably led to strong profit growth.

The role of the Government in this skewed distribution of benefits should not be underestimated. It was Government which imposed the coal contracts on generators and RECs in 1990: this directly led to a structure where, for the 1990 to 1993 period, domestic electricity consumers had to bear the burden of the extra coal costs. Equally important is that the initial price-cap settlements for the monopoly parts of the industry were determined in 1989/90 by the Government and not by OFFER, and it was not until 1995 – in the distribution price control review – that OFFER had a chance to have a real impact on the prices charged in the monopoly parts of the industry. OFFER's new price controls, set between 1993 and 1995, have not yet had much effect in restraining profit growth and probably still favour shareholders unduly. However, they are just beginning – together with much lower coal prices – to produce some real price reductions for consumers, especially domestic consumers.

Part III

THE ISSUES

10

Competition: The Continuing Issues

Mike Parker

THE FRAMEWORK OF COMPETITION

For England and Wales, which make up the vast majority of the UK electricity supply industry, a major feature of privatization was the separation of supply, distribution, transmission and generation into separate businesses, although supply and distribution both stayed within the ownership of the 12 RECs. The main purpose of this reorganization was to separate the management of those activities which were effectively 'natural monopolies' (distribution and transmission) from supply and generation which could be made competitive. It was recognized that the 'natural monopolies' would require regulation for the foreseeable future; but for supply and generation, the regulator was charged among other things with promoting competition. For these activities any regulation was regarded as merely transitional, until such times as full and effective competition had been established.

Excluding the fossil fuel levy, an impost due to be phased out following the planned privatization of Nuclear Electric, potentially competitive activities make up on average 70 per cent of the total cost of electricity to final consumers, although more to large users and less to small users (Table 10.1).

Prior to privatization, in England and Wales the CEGB – subsequently divided into National Power, PowerGen, Nuclear Electric and the National

Table 10.1 Analysis of costs to final electricity consumers (user's level of consumption) (%)

	0–100kW	*100kW–1MW*	*Over 1MW*	*All users*
Potentially competitive				
Generation	58	66	77	65
Supply	8	4	1	5
Subtotal	*66*	*70*	*78*	*70*
Natural monopolies				
Transmission	5	6	5	5
Distribution	29	24	17	25
Subtotal	*34*	*30*	*22*	*30*
Grand Total	*100*	*100*	*100*	*100*

Source: Trade and Industry Select Committee (TISC): Aspects of the Electricity Supply Industry (HC481-1 July 1995). Adapted by author.

Grid Company – had a monopoly of both generation and transmission. The local area boards, which became RECs, had the monopoly of local distribution and supply. By the end of the first five years of privatization, these dominant positions had been significantly eroded (Table 10.2).

The RECs' franchise monopoly of supply for customers using less than 100 kW per year, mainly the domestic market, is due to terminate in 1998. The franchise for the 100 kW–1MW market was opened to competition in 1994/95, and in early 1996, this market is still developing.

In terms of the numbers of sites supplied, the position is very different (Table 10.3). In those supply markets which had been opened to competition, the RECs saw a considerable erosion of their markets by 1994/95. The market share lost by the 'local' RECs in the five years was 50 per cent in terms of electricity supplied and 26 per cent in terms of number of sites. The market share lost by the RECs collectively in terms of electricity supplied was 33 per cent, but only 7 per cent in terms of sites supplied.

Thus, although the 'local' RECs and the RECs collectively, have seen significant erosion of their market share in terms of electricity supplied in those market sectors that have been opened to competition, this change has been disproportionately concentrated on the larger industrial and commercial sites. (The domestic and small-commercial market, accounting

Table 10.2 Erosion of 'local' RECs' market shares 1990–1994 (user's level of consumption) (%)

	100 kW or less		*100 kW to 1MW*		*Over 1 MW*	
Percentage of total supply market	*50*		*20*		*30*	
Market shares	1990/91	1994/95	1990/91	1994/95[1]	1990/91	1994/95[1]
REC first tier	100	100	100	70	57	37
REC second tier	–	–	–	21	4	15
Major generators and others	–	–	–	9	39	48

Source: OFFER
1 OFFER Estimate (1994 Report).
REC 'first tier' supplies are those within the area of its distribution monopoly. 'second tier' are those supplies outside that area.

Table 10.3 Erosion of 'local' RECs' market by site 1990–1994 (user's level of consumption) (%)

	100 kW or less		*100 kW to 1MW*		*Over 1 MW*	
Number of sites supplied	22 million		45,000		5,000	
Market shares	1990/91	1994/95	1990/91	1994/95[1]	1990/91	1994/95[1]
REC first tier	100	100	100	76	72	55
REC second tier	–	–	–	19	4	21
Major generators and others	–	–	–	5	24	24

Source: OFFER
1 OFFER Estimate (1994 Report)

for half the total UK public-supply market still remains subject to the monopoly supply franchise of the RECs until 1998.)

It needs to be emphasized that the above developments in supply competition have no effect on the RECs' monopoly of distribution (the local 'wires business') which remains overwhelmingly their main source of profit, yielding a substantial profit margin. By contrast, the supply business is one of large turnover and small margins. The figures for the RECs collectively in 1993/94 are shown in Table 10.4.

Table 10.4 Profitability of RECs' distribution and supply business, 1993/94

	Turnover (£m)	*Operating Profit (£m)*[1]	*Margin (%)*
Distribution	3,928	1,083	28
Supply	13,723	237	2

Source: OFFER
1 Current cost accounting (CCA) basis.

The position of competition in generation is rather different. The market share of the main incumbent generators, National Power and PowerGen (who inherited all the fossil fuel plant of the former CEGB) has been eroded by that part of the 'dash for gas' undertaken by independent power producers (IPPs) and by the improved performance of Nuclear Electric's stations. The change over the period to 1994/95 is shown in Table 10.5.

Further IPP gas stations already committed, together with the regulator's requirement that National Power and PowerGen should divest 4 GW and 2 GW of plant respectively could mean that their collective market share could be further reduced by up to 20 per cent by the end of the decade. However, in spite of the significant reduction of the market shares of the dominant generators, the profitability of their generation has increased. The collective operating profit (CCA basis) margin on turnover of National Power and PowerGen increased from 5 per cent in 1990/91 to 15 per cent in 1993/94 (the position of both companies being similar). Most of this profit has come from the 'back-to-back' contracts between the generators and the RECs and between the generators and British Coal (and privatized successors) which, in spite of the high cost of British Coal, embodied a substantial profit margin.

The objective of introducing competition into electricity is to benefit consumers and the economy generally by reducing costs and prices. Basically, this can be obtained in two ways: first, by reducing resource costs per unit of production, and, second, by reducing profit margins, where these are larger than would normally be expected in competitive condi-

Table 10.5 Major generators' shares of output (%)

	1989/90	*1994/95*
National Power	48	34
PowerGen	30	26
Nuclear Electric	16	22
Others	6	18
Total	*100*	*100*

Source: Trade and Industry Select Committee (TISC)

tions. The above preliminary analysis indicates that, in terms of the overall potential benefit to consumers, effective competition in generation is many times more important than competition in supply, since generation constitutes 65 per cent of total electricity costs as against 5 per cent for supply costs. This is not to say that supply competition is of negligible importance. The extension of supply competition to the domestic market needs to be considered not only because (it is claimed) competition in generation cannot function satisfactorily unless there is supply competition in all markets but also because it will raise a number of potentially politically-sensitive issues concerned with 'social obligations' and 'winners and losers'. Finally, consideration has to be given to potential changes in the ownership structure of the electricity industry in terms of mergers and takeovers and the effect this might have on the nature of competition. These matters are dealt with in turn below.

Competition in Generation

There has been much discussion on the adequacy or otherwise of competition in generation in England and Wales. The position in Scotland and Northern Ireland is different. But there has been less debate on the characteristics of a fully competitive market towards which policy might aspire. These characteristics are utilization of existing plant determined by the short-run (daily) avoidable costs of the marginal plant required to meet demand day-by-day; availability of existing plant determined by the short-run (annual) avoidable costs of the marginal plant required to meet peak demand over the year; free entry of new capacity on the basis of 'net effective cost' to the system, either to meet incremental demand or to replace existing plant, using a discount rate reflecting the generators' cost of capital; and sufficient diversity of ownership to ensure that there is downward pressure on generation prices across the range of load-factors from base load to peak generation.

Some six years after privatization, there is still significant divergence from these characteristics of a fully competitive market. The emphasis of regulation has been on reducing the dominant market share of the two major generators, National Power and PowerGen. As we have seen, this had been eroded by 16 percentage points by 1994/95, with a similar further fall likely by the end of the decade. But this loss of market share has been much more pronounced in the case of base-load generation, where the share of the two major generators had by 1993/94 fallen to only 55 per cent. The reasons for this are twofold. Firstly, the increase in the output from Nuclear Electric stations has automatically added to base-load supply (since nuclear stations have low avoidable costs, and cannot readily 'follow' load) – a process reinforced by the requirement under the Pool rules that available nuclear plant 'must run'. Secondly, the introduction of new CCGT plant by IPPs (the 'dash for gas' – see Chapter 6) was generally associated with long-term contracts (typically 15 years) for the supply of electricity to RECs (normally partners) and for the supply of gas, with a high level of take-or-pay. These new entrants were encouraged by the regulator.

The two major generators have claimed that the introduction of competition in this way has represented a market distortion, particularly in the lucrative base-load market, which has reduced the load on their coal-fired stations, even though their avoidable costs were lower than the total costs (including capital charges) of many of the new CCGTs. In terms of the continuing reduction in the market shares of the dominant generators there appears to have been a significant increase in plant ownership diversity, but the mechanism by which this has been achieved represents in large measure a market distortion.

Because new entrants have been concentrated in the base-load market, the dominance of National Power and PowerGen in setting Pool prices has not been reduced. Mid-merit and peaking plant owned by the two generators has effectively set the Pool prices for nearly 90 per cent of the time, the balance being set by pumped storage capacity. IPPs have had little influence on Pool prices, and, because of the nature of their take-or-pay contracts, have had to bid into the Pool at whatever low price was necessary to ensure base-load generation. Such low bidding to secure load carries no threat, since generators are remunerated according to the strike price in their 'contracts for differences' and not according to their bid prices.

There has always been competition between individual stations to secure load in an integrated system. The CEGB's merit order operation was effectively a competitive market to determine the load of individual stations based on their avoidable costs. Such competition is inherent in any system when most plant is technically capable of operating at or near base-load and where the average plant utilization is only about 50 per cent (because of the load profile of demand for electricity, which cannot be stored). However, the CEGB as a whole was under no commercial pressure to reduce costs, or its bulk supply tariff, because its monopoly position of supplier to the Area Boards, which were themselves monopolies, allowed total pass-through of costs.

Since privatization, all generators have had commercial incentives to reduce costs, in order to maximize profits at any given level of Pool or contract prices for wholesale electricity. The more difficult issue is to establish how these cost savings can be shared with final consumers (as would normally be the case in competitive markets). This would appear to require effective competition in non-base-load generation – that is, at the margin where Pool prices are set, and which therefore determine in large measure the wholesale electricity price which constitutes some two-thirds of the price to final consumers.

There are considerable obstacles to achieving this. Above all, the inheritance of the initial ESI privatization structure which established the duopoly of National Power and PowerGen is difficult to change except by erosion over a long period. Each of the three possible routes to creating effective competition in mid-merit and peaking plant is difficult.

Firstly, it is improbable that new entrants would seek to build new plant with the express purpose of operating at medium-to-low load factors. In order to provide a return on capital, new entrants have sought to operate at or near base-load, and entry by new capacity on any other basis is less likely to be financially viable. Indeed, nearly all existing plant now working intermittently was originally constructed to serve base-load.

Secondly, the 'trickle-down' of IPP plant into mid-merit operation would require a substantial surplus of base-load plant not owned by National Power or PowerGen. Leaving aside available nuclear plant, which will stay on base-load until it retires, existing IPP plants will be under contract for at least a decade ahead, under terms which make it unlikely that they will voluntarily reduce load, except where there might be profitable opportunities to sell-on gas under arbitrage arrangements. Additional IPP plant might be built without long-term contract cover, but investment in such plant may well not be prudent if there were a real chance that competition would quickly force them down the merit order into medium load. Any indication that public policy favoured such an outcome would represent a major disincentive to new entry.

Thirdly, there is little possibility that National Power and PowerGen will compete with each other in such a way as to affect Pool prices. Although electricity demand is insensitive to price in the short run, demand for a particular generator's electricity is very sensitive to price. Any price reduction by one generator would therefore immediately have to be matched by the other. Both have a strong incentive to avoid this outcome and, given the duopoly, they are in a strong position to do so.

The only way to get effective competition in mid-merit plant within the foreseeable future is therefore to create a diversity of ownership of such plant. This involves divestment of plant by National Power and PowerGen. The regulator has required the divestment of 6 GW plant for this reason. Yet the technical problem remains: how to ensure that the new owners operate the divested plant at sufficiently low load factors as to have an influence on Pool prices, even though there would be strong financial incentives to operate at higher load factors. At the time of writing (May 1996), it is not clear whether the conditions of sale or new SO^2 emission limits will limit the load factor of the divested plant in an effective way. Even if this difficulty is overcome, National Power and PowerGen would still account for about a 50 per cent share of price-setting in the Pool, and it is by no means clear whether one or two new entrants would change the inherently collusive nature of the price-setting process. Alterations to the Pool have been proposed as a means of securing more effective competition in generation. All these could involve some difficulty.

There have been proposals that generators should be able to make bilateral deals with consumers outside the Pool. It has been claimed that such arrangements could give rise to selected customers (probably large energy-users) obtaining special deals which failed to recover in full the costs of running a secure integrated system from which all customers benefit. The primary purpose of the Pool is to secure a merit order of operation while at the same time delivering market clearing prices for wholesale electricity which cover the costs necessary to achieve a high degree of security of supply for the system as a whole since these costs are difficult to allocate; trading outside the Pool could give rise to a 'free-rider' problem.

There have been suggestions that generators should be remunerated in line with their bids rather than by Pool price or contracts related to Pool price, but this would compress price differentials between peak and off-peak, which would distort the market. There have been proposals to extend demand-side bidding into the Pool, but the difficulty is that customers would have no contractual obligation to take forecast demand.

Any radical change to the Pool would be difficult, not least because the Pool is not a corporate body and has to work through agreement among all its members. At April 1995, it had 50 members (27 generators and 23 suppliers) who are signatories to a contract – The Pooling and Settlement Agreement – which is not easily amended.

It has been argued that 'full' competition in supply following the end of the RECs' franchises in 1998 will itself provide a boost to effective competition in generation since it would no longer be possible for a proportion of generating costs (where these were higher than justified by free markets) to be passed through into a captive franchise market. But while supply competition may well be an important stimulus to effective competition in generation, it would not, of itself solve the problem of the dominant position of National Power and PowerGen in the price-setting mechanism. Further, even if effective competition in generation and in supply can be established, there are still difficulties in determining how this would flow through to give lower prices to final consumers than would otherwise have been the case. If in fact genuine competition is established in mid- and low-merit plant then the companies concerned would have a further incentive to reduce the avoidable costs of such plant, and would be obliged to accept lower margins than hitherto. This should reduce the system marginal price in the Pool, although it is not clear whether such reduction might be offset to some extent by higher capacity charges (arising from the operation of the LOLP x VOLL formula).

A more fundamental point is that the Pool price will dominate the wholesale price of electricity after full liberalization in 1998, even if most of the generators' output continues to be sold under contract. On the assumptions that there is no significant trading outside the Pool, that dispatch of the output of individual stations continues to be determined by marginal bids, and that these marginal bids continue to set the system marginal price, which is the main component of the Pool purchase price, contracts for the sale of generated electricity will rarely if ever carry an obligation to supply or take, but will continue to be in the form of two-way 'contracts for differences' (CFDs) embodying a strike price, as a financial instrument to hedge price risk under conditions of volatility. Further, given that the output or market share of any generator is determined in the Pool by the same process that determines the system marginal price and given the short-term price inelasticity of total electricity demand, no generator would seek to obtain higher sales or market share by selling at discounted Pool prices. Such a course would merely reduce revenue for the same volume.

In these circumstances, it is likely that once the RECs' franchise and the former BC coal contracts come to an end in 1998, there will be broad convergence between CFD strike prices and Pool Prices on an annual basis, and a trend towards short-term (one- or two-year) contracts linked to expected Pool prices, with any modest contract premium reflecting the wish of most customers to 'hedge' the risks of uncertain and volatile pool prices. As a consequence, the primary competitive vehicle in which wholesale electricity price reduction was achieved would not be price differentials in contract offers by the various generators, but rather the effects of competition in reducing the Pool price which would continue to act as the reference point for CFD prices, even if (as now) nearly all electricity was sold under contract. Pool price-setting would thus be central to the issue of securing benefits from competition.

In addition to the issues of effective competition between existing plant, there is also the question of whether the system provides adequate competition in the provision of new plant. We have seen that, since privatization, the provision of new plant (as part of the 'dash for gas') has been the result of other than pure market forces. Moreover, the mechanism in the Pool to provide capacity payments (LOLP x VOLL) has so far had no impact as a market mechanism for new investment. VOLL (value of lost load) itself is an arbitrary judgement, and capacity payments arising from the formula have been very volatile – on occasions involving extreme price 'spikes' – but on average, insufficient in themselves to trigger new investment. The Association of Independent Electricity Producers in their evidence to the TISC 1995 enquiry stated that since 1990 Pool prices had been insufficient to remunerate new investment (although this had not inhibited the 'dash for gas'). In the longer term, the full liberalization of the supply market in 1998 is expected to make suppliers averse to long-term contracts (although short-term CFDs will continue to be used to hedge short-term variability in Pool prices). This situation could be reversed if, unlike recently, there were a strong upward trend in gas and coal prices, but the probability that a competitive market will discourage long-term contracts could act as a major disincentive to investment in new capacity – unless such capacity is substantially profitable after taking account of capital charges. Under fully competitive conditions, the risk-aversion of generators could, in the long run, prove to be incompatible with the high degree of system security demanded by the public.

We conclude that although the market share of the two major generators, inherited from the original privatization settlement, is being substantially reduced, this has been achieved in some degree by a market distortion and has not led to effective competition in the determination of the wholesale price of electricity (as distinct from reductions in the operating costs of generation). There are great difficulties involved in rectifying this other than over a period of years. In addition, full competitive mechanisms to determine adequate new investment in plant have yet to emerge.

COMPETITION IN SUPPLY

As we have shown, the link between the introduction of full 'supply' competition and the establishment of effective competition in generation is by no means clear. We now look at the effects of supply competition in isolation from the question of generation.

We have seen that supply costs are a negligible proportion of total costs of providing electricity to large consumers, so that there can be no significant benefit in terms of price reduction arising from competition in these particular costs. Even with domestic consumers, where the supply costs form a larger proportion of total costs, the potential benefit in 1998 is small. For example, if competition were to halve these costs, then prices would be reduced by only 3-4 per cent – say £10 on an average annual domestic bill of about £260. Yet this could well be optimistic, given the small margins (currently 2 per cent of turnover) available on the supply business. For the average domestic consumer, savings due to competition in supply, excluding the possible effect of such competition on the wholesale price of electricity through the stimulation of competition in generation, would be strictly marginal. There might be other benefits, such as the ability of suppliers to offer electricity and gas together, energy efficiency packages and so on, but the monetary value of these benefits will be small for the average consumer.

To set against these small price benefits, there are the costs of introducing a new metering and settlement system to enable full cost-reflective pricing on a half-hourly basis to be introduced. High metering and settlement costs have already been seen in the 100 kW–1 MW market, where competition was introduced in 1994. It has been estimated (by TISC) that the extra cost of introducing 'smart' meters in the domestic market could be £1 billion, or at least £50 per customer, an outlay which would not appear to be justified in terms of the savings in supply costs alone for the average domestic consumer – although in the longer term, sophisticated cost-reflective metering could lead to reductions in costs for the system as a whole.[1]

An even greater problem is the sheer complexity of the settlement system which would ensue. The number of sites supplied in the under-100 kW market is over 400 times greater than the number of sites supplied to these market sectors where supply has already been liberalized. The complexity of the settlement system would be further increased if more sophisticated meters were introduced. At the time of writing (early 1996), it is proposed to mitigate this problem of complexity by introducing 'profiling', where the profile of use for particular socio-economic groups of customers would be estimated as a surrogate for half-hourly metering. It is not clear whether this will be a satisfactory or equitable substitute for individual customers, nor quite how it would work in practice. Could individual customers demand to be transferred to another profile if they thought they were being unfairly treated? Could competing suppliers attract more customers by offering to transfer them to more advantageous profiles? If the profiles were sacrosanct who would ensure that millions of customers were in the right profiles?

Overall, therefore, it is doubtful whether there will be benefit to the domestic market as a whole, from supply liberalization, unless this leads to lower prices through greater competition in generation.

The extension of supply liberalization to all markets by 1998 will raise a number of issues of cross-subsidy. 'Full' competition tends to drive out cross-subsidy but in the electricity industry this does not necessarily apply, since a substantial part of total costs will continue to reside in the natural monopoly activities of transmission and distribution where, by definition, competition can have no impact. Here, the extent to which cross-subsidies between and within market sectors are identified and eliminated will depend on the policy on regulation.

To date little evidence has been put forward which would identify cross-subsidy between, say, large industry and the domestic market, or between large and small domestic consumers, or between regions. Some have suggested that this problem may not be so great as was thought to be the case with gas, as regional electricity price differentials are already significant, and there are already fairly wide differences between prices paid by large and small consumers. Nevertheless the chances are that a competitive supply market would rapidly be forced to a charging system in which fixed costs were recovered by a 'capacity charge' with the balance through an 'energy' charge: large customers would gain, and others lose. As the majority of domestic consumers take less than the average, the results would be unpopular. The 'winners' and 'losers' issue has proved to be politically sensitive in the preparations for competition in the domestic gas market, which has required several special measures.

On the basis of current charges, larger and better-off domestic customers are likely generally to be more profitable to supply than smaller customers (often the socially disadvantaged). Hence, there will be a tendency for new entrant suppliers to 'cherry-pick' these more profitable customers, leaving the less profitable customers and thus a disproportionate share of the 'social obligations', with the incumbent supplier (in the case of the domestic market, the local REC). Potential new entrants have advocated that this should be 'left to the market', in order to facilitate the introduction of competition. Not surprisingly, RECs have stressed the need for measures to ensure that any burden of 'social obligations' should be equally shared from the outset. Given the political sensitivity, the gas regulator has been given powers under the 1995 Gas Act to impose a levy on suppliers with a smaller-than-average ratio of customers who are classified as 'social obligations'. It remains to be seen whether similar measures will be introduced when the domestic electricity market is liberalized.

The introduction of competition into the supply of domestic electricity is almost certain to create 'winners' and 'losers', though this might be mitigated by licence conditions and by regulatory action on transmission and distribution charges. The political acceptability of the whole process will greatly depend upon whether the average level of electricity prices is falling – in large measure dependent on the trends in fossil fuel prices and the rate of general inflation. If fossil fuel prices fall – which has generally been the case in the 1990–1995 period – and inflation is low, then the absolute increases in prices to the 'losers' will be at worst modest, and the reductions for the 'winners' will be substantial. On the other hand, if fossil fuel prices rise and inflation rates increase, the 'losers' would face significant (and unpopular) increases, and even the 'winners' would appear to see little advantage.

There is also the question of whether the high level of supply security can be maintained with full supply competition (over and above the issue of the adequacy of the VOLL x LOLP formula in triggering new generating investment – see above). It is by no means clear how the totality of supply contracts will match demand, or whether a 'supplier of last resort' will be required. Moreover, under a competitive system, no one supplier has an interest in maintaining sufficient spare capacity to maintain system security, which becomes an 'externality' so far as individual players in the market are concerned.

Whether or not some of the concerns expressed prior to liberalization of domestic electricity supply prove to be ill-founded, it is clearly a task of enormous complexity and logistical difficulty, with a whole raft of associated issues having a substantial political content. In these circumstances, the effective management of the process of introducing competition is vital to its success or, to put it negatively, to prevent disaster. In their report 'Aspects of the Electricity Supply Industry' (HC481 July 1995) the Trade and Industry Committee expressed great concern as to the adequacy of the planning, particularly as it is proposed to move, at one step, from a franchised market to a fully liberalized market in April 1998. On this issue the Committee recommended that

1 'The Government ensures that, for the specific task of introducing full competition in supply, there is a central coordinator – with sufficient powers to organise the process and take responsibility for its success or failure' (para 25).
2 The Regulator should 'set out a timetable for producing urgently a workable system of profiling' (para 29).
3 'There be trials of full competition in supply [ie before 1998]... involving a significant number of customers' (para 30).

Subsequent responses by the Government and OFFER indicate that these organizational issues are being taken seriously. The regulator is taking the lead in ensuring that the 1998 liberalization is implemented effectively and on time.

COMPETITION AND INDUSTRY RESTRUCTURING

Under the original privatization settlement, the Government held 'golden shares' in the RECs until April 1995. This effectively precluded changes in the structure of the industry arising from mergers or takeovers until after that date. In fact the first bid, by the conglomerate Trafalgar House for Northern Electric, was made several months before the 'golden shares' lapsed; but although this bid was subsequently withdrawn, there was subsequently intense takeover activity. Table 10.6 summarizes the bids which had been made by December 1995.

Table 10.6 Takeover bids for RECs, 1995

REC	*Bidder*	*Status (at January 1996)*
Seeboard	Central and S West (US utility)	Agreed[1] and approved[2]
South Western	Southern Company (US utility)	Agreed and approved
Norweb	North West Water	Agreed and approved
South Wales	Welsh Water	Agreed and approved
Eastern	Hanson (Conglomerate)	Agreed and approved
Manweb	Scottish Power	Agreed and approved
Midlands	PowerGen	Agreed but referred to MMC
Southern	National Power	Agreed but referred to MMC

1 'Agreed' means agreed between the parties; 'approved' means approved by DTI.
2 The National Power and PowerGen bids lapse while the MMC investigates.
3 As at January 1996 only 4 of the 12 RECs were not subject of bids – Northern, London, Yorkshire and E Midlands.

Within a year of the end of the 'golden share' arrangement lapsing, 6 of 12 RECs have ceased independent existence. Clearly, this degree of restructuring raises a number of policy issues. In August 1995, the Government, having at that time approved three takeovers without reference to the MMC, stated that its policy was that the RECs should be subject to the normal opportunities in the market place, including involvement in mergers and acquisitions, although it recognized that such developments in the electricity industry might raise special issues.

The special issues fall into two categories; regulatory concerns where the regulator might find it more difficult to carry out his or her duties following mergers or takeovers; and whether an enhanced degree of vertical integration might reduce effective competition. The potential regulatory effects are dealt with in Chapter 11. We deal here with the extent to which takeovers and mergers might inhibit the benefits from competition.

If there were 'horizontal' mergers of RECs to an extent which greatly reduced the number of suppliers to final markets (particularly the domestic market), this might be seen as anti-competitive. However, by the end of 1995 no such horizontal mergers had taken place. Mergers between RECs and water companies, and takeovers by US utilities and/or conglomerates do not appear to raise competition issues, provided the regulatory problems of reduced transparency can be overcome.

The real issue concerns vertical integration. Two of the takeovers approved during 1995 might potentially raise this issue. Scottish Power has a vertically integrated business, but its takeover of MANWEB was allowed on the basis that no special deals would be made for power generated in Scotland. Hanson's takeover of Eastern might raise vertical integration issues in the light of its (planned) acquisition of 6 GW of divested plant from National Power and PowerGen. However, the vertical integration issue was brought to the fore by the bids by National Power and PowerGen for Southern and Midland RECs respectively. The Government's response was to refer both bids to the MMC. The Minister stated that:

> 'In general I do not believe that vertical integration is inherently objectionable whether in the electricity industry or elsewhere. However, in these two cases, the structural change proposed could have an effect on the development of competition in the industry'.[2]

In April 1996, it was reported[3] that the MMC had recommended that the bids by National Power and PowerGen for Southern and Midlands Electricity respectively should be allowed, subject to relatively modest conditions. In the event, the Government decided not to allow the National Power and PowerGen bids for the two RECs to proceed. The Minister stated that,[4] although he remained of the view that vertical integration was not inherently objectionable, 'in the current state of the market, there would be significant detriments to competition if these mergers proceed', which the remedies proposed by the MMC were insufficient to overcome. Yet this leaves the future of vertical integration unclear, since the RECs can still increase vertical integration by acquiring generating plant, as Hanson's Eastern Group is planning to do.

The question of whether vertical integration is anti-competitive or otherwise undesirable is by no means clear-cut. From 1998, it is proposed that the RECs will lose their monopoly franchises in the domestic market (and are already subject to competition in other markets). Provided that effective competition in supply obtains in all markets and that effective competition in generation can be established, then a generator/REC merger would not confer any anti-competitive advantage to the generator in a way which would allow higher costs to be passed through; and, given that the RECs' distribution business would continue to be separately and effectively regulated, there is no inherent reason why an REC/generator merger should be anti-competitive. Any attempt to pass through higher costs would damage the competitive position of the merged company in the final market.

But if it proves difficult (for the reasons outlined above) to establish fully effective competition in generation, and if competition in the domestic market is slow to emerge, then vertical integration will continue to raise anti-competitive issues – at least for a further transitional period.

In any event, given the higher risks to generation arising from competition, and the small margins available on 'supply', there will always be a tendency for companies to seek to reduce these risks by integration and/or long-term contracts. It will continue to be necessary for Government to have a clear view on the extent to which vertical integration is to be permitted, and under what conditions.

Before the Government had made its decision on the MMC reports, the position had been further complicated by the announcement by the US utility, Southern Company, that it might bid for National Power. In May 1996, the Government stated that it would retain its 'golden shares' in the two major generators, at least for the timebeing, 'in view of the importance of National Power and PowerGen as independent generating companies operating in a market which is not yet fully competitive'.[5] Foreign ownership of the ESI (particularly by US utilities) was beginning to emerge as a political issue.

There is a further uncertainty: namely what new suppliers will enter the market when the domestic market is opened to competition. We have seen that the supply business is one of high turnover but low margins. This suggests that 'stand-alone' supply businesses will be unlikely: rather, supply businesses will form part of larger companies – with links to other parts of the ESI, such as major fuel companies (eg oil companies), or major retailing companies (eg supermarket chains).

CONCLUSIONS

Four broad conclusions can be drawn from the above discussion of the issues arising from the introduction of competition into the electricity industry. Firstly, competition in 'supply' is easiest for large industrial customers, and most difficult for the domestic market. Indeed, the sheer complexity and transaction costs involved, together with the small proportion of final prices represented by supply costs and the difficult political issues of 'winners' and 'losers' are likely to make domestic-market liberalization in 1998 of dubious value, *unless* it has the effect of creating more effective competition in generation than would otherwise have been secured.

Secondly, the key to providing price advantages to final consumers (whether domestic, commercial or industrial) from competition lies overwhelmingly in generation, since the other components of cost are either small, in the case of supply, or regulated, in the case of distribution and transmission. With generation, the issue is how to transfer a larger share of cost savings (the incentives for which already exist) to *price* reductions to final consumers by reducing the 'wholesale' price of electricity. In this regard, the main obstacles derive, either directly or indirectly, from the original privatization structure and settlement. The removal of these inherited obstacles will be a both difficult and lengthy process.

Thirdly, once takeovers and mergers were allowed, the initial privatization structure was subjected to rapid and radical challenge, which raised a further raft of issues relating to the model of competition being adopted, as well as potential regulatory problems. Given the uncertainty as to what new entrants there will be, and what the corporate strategies of the various companies will be, the future shape of the industry cannot be predicted. The main policy issue is the extent to which vertical integration is to be regarded as anti-competitive.

Fourthly, it is by no means clear whether the establishment of fully effective competition in generation and supply will be compatible with the degree of long-term system supply-security demanded by the public, since competition in generation will increase generators' cost of capital, and individual suppliers will have no commercial interest in the security of the system as a whole.

11

Unresolved Issues of Economic Regulation

John Surrey

Privatization involved bold changes in the structure and economic regulation of the electricity supply industry, and it was perhaps inevitable that the regulation would evolve as the new structure settled down and experience built up. But after five years' experience, there is increasing awareness of deficiencies and unresolved issues and questioning of certain aspects of the British approach to economic regulation of the utility industries.

This chapter discusses the unresolved issues and deficiencies of the British approach to regulation and how these should be dealt with. There are three main sections. The first is on the structure and accountability of the economic regulation. The rationale for this is that inappropriate institutional arrangements may result in poor regulatory decisions and insufficient opportunity to appeal against them, and that inadequate public accountability could reduce public acceptability. The second section is on the problems of regulating the natural monopoly network consisting of the national transmission grid and the regional distribution networks. These are the parts of the industry which, as was always evident, will need continuing regulation even if generation and 'supply' are eventually fully competitive on a sustainable basis. This section also contains a discussion of alternatives to price regulation which might produce results that are more publicly acceptable, and of some problems concerning the regulatory treatment of capital and investment which are likely to assume greater importance. The third section, on 'Other Regulatory Issues', discusses the regulatory issues posed by takeovers and mergers of the RECs, and of the problems of achieving competition in generation and 'supply'.

Before proceeding, it is useful to have in mind the objectives of economic regulation.[1] A working definition is that economic regulation should achieve a satisfactory balance between the interests of consumers, shareholders and other stakeholders (managers and employees); that it should secure sufficient investment to meet present and future demand and maintain a high quality of service at all times; that it should provide incentives for internal cost-efficiency and for innovation; and that it should also take account of relevant wider objectives – including the environment and energy efficiency, in the case of electricity. Together with the privatization statutes, the prospectus, and the supply licences, economic regulation forms a 'social contract' which aims to retain the confidence of both consumers and shareholders in the performance of the utility companies in terms of service standards and the distribution of the economic rent between lower prices for consumers on the one hand and higher dividends for shareholders on the other hand. Such a balance is unlikely to remain static for long. Rather it is likely to change through time in response to pressures from either side due to changing events and perceptions of risks.

The British approach to economic regulation in the electricity and other utility industries is not in crisis or near breakdown. But the regulation has been repeatedly criticized for the large profits it has allowed on the low-risk monopoly activities of transmission and distribution, for the large boardroom pay increases and stock options, and for the meagre benefits so far to domestic electricity consumers despite the large fall in primary fuel costs and the continued improvement in labour productivity which began under public ownership (see Chapter 2). Other concerns have been longer-term supply security, the failure to take account of relevant wider policy objectives such as energy efficiency and R&D, the problems raised by the takeover activity in the electric power sector in 1995/96, and the lack of preparation made by the industry up to early 1996 for the transition to competition in 'supply' in the domestic market in 1998. These criticisms and concerns have tended to erode the public acceptability of the regulation. The failure to control profits adequately was behind the revised 1995 distribution price control, the need for which was revealed by the unsuspected cash hoard which Northern Electric used to fight the Trafalgar House bid.

By no means all the deficiencies were the fault of the system of regulation as such, nor will they necessarily be long-lasting. The large profits of the National Grid Company (NGC) and the RECs over the first five years, and the RECs' substantial windfall gain from the flotation of the NGC, were due to the Government's generosity in setting the initial transmission and distribution price controls and its undervaluation of the transmission system. The level of dissatisfaction with electricity prices and profits could subside now that customers have each been given £50 from the proceeds of the flotation of the NGC, as considerably lower domestic electricity prices result from the 1995 distribution price control, and also if the company directors limit their own pay increases as strictly as they limit their employees' pay increases.

Although this chapter will focus on the role of the electricity regulator, it is important to recognize that economic regulation of the electricity supply industry can be influenced from time to time by the Director General of the Office of Fair Trading (OFT), the Monopolies and Mergers Commission (MMC), the Department of Trade and Industry (DTI) and the European Commission (EC). The DTI, as the Department responsible for competition policy and policy for the utility industries, has a close interest in the direction of economic regulation and the promotion of competition in the utility industries and, through its Secretary of State, appoints the electricity regulator. So far the EC has generally been on the sidelines of British economic regulation of the utility industries given that Britain is much farther down the path of liberalization which the EC wants all other member states to tread. Assuming that further European integration takes place and that Britain will be included, the EC could eventually have greater importance with regard to economic regulation in Britain.

The Structure and Accountability of Economic Regulation

Various criticisms have been made from the outset of the structure and accountability of the British system of economic regulation of the utility industries. These criticisms have been captured under the much repeated headline of 'Who regulates the regulators?' There are four main criticisms.

It is claimed firstly that the system of regulation is administered by officials appointed by the Government and that there is insufficient accountability to Parliament on behalf of the public for decisions which can have far-reaching economic and social effects. Secondly, it is claimed that the system gives considerable scope for discretion in regulatory decisions, gives rise to idiosyncratic, personalized styles of regulation, does not require detailed explanation or justification of the decisions, and that these factors make it difficult effectively to challenge the regulators' decisions.

Thirdly, it is argued that, although in principle the regulators are independent, in practice they are too close to government by virtue of the fact they are appointed by the relevant Minister and have to work within a framework of policy laid down primarily by the Government. Compared with the period of public ownership, the Government is no longer in the driving seat as far as the utility industries are concerned, but it is widely thought to be the back-seat driver steering the economic regulators on all important matters. Finally, the economic regulation of the privatized utilities has so far failed to reflect other wider policy objectives, including, especially, efficient energy use.

PROPOSED STRUCTURES

Before discussing the reforms which have been proposed to deal with these criticisms, it is worth noting two points. Firstly, there is no long tradition in Britain of 'one-person' regulators. (By 'one-person' regulators we mean, for example, the Director General of Electricity Supply who is statutorily responsible for all economic regulation of electricity supply; but he is, of course supported by a substantial staff at OFFER, which is purely advisory.) The idea seems to have arisen with the decision in 1983/84 to privatize British Telecommunications when the Government saw the need for regulation as purely temporary and there is no indication that consideration was given to the long-term requirements of the task. There is, however, considerable similarity between the 'one-person' economic regulators and the chief executive officers of the agencies now responsible for the implementation of government policy in many fields. Ministers can now claim they are responsible for setting policy but not for implementing it. Ministers remain the power behind the scenes, but the blame when things go wrong tends to get located with the 'implementers' as opposed to the policy-makers.

Secondly, all the proposed reforms involve only a reorganization of responsibilities. Irrespective of whether such changes are necessary or desirable, it is important that the number and expertise of the regulatory staff employed match the requirements of the task. This is likely to require recruiting highly qualified and experienced staff and ending the dependence on both civil servants on temporary attachment and on outside advisers and consultants who bear no responsibility for the outcomes.

A Regulatory Commission for each Industry

Replacing the 'one-person' regulators with commissions with three or five members would overcome the main criticisms of 'one-person' regulators. Commissions would neutralize the personality factor, produce more stable regulation, and diffuse the media pressure among several commissioners. The commissions would be based on the principle that several heads are likely to be wiser than one (as with magistrates' courts, trials by jury and courts of appeal). With commissions the responsibility is shared and the decisions are based on a majority view. Commissions, by the way, are not to be confused with the wish of some of the present 'one-person' regulators to appoint panels of advisers whose advice could be accepted or rejected, leaving the Director General still solely responsible for the decisions. If 'one-industry' regulatory bodies are to be retained, there is a good case for switching from 'one-person' regulators to regulatory commissions. We next examine the alternatives to 'one-industry' regulatory bodies.

A Regulatory Commission Covering all the Utility Industries

Should the present 'one-industry' regulatory bodies be replaced with a single Regulatory Commission covering all the privatized utility industries? The single Commission would be close to the US model of state-level public utility commissions. Such a body could have several advantages, including more uniform treatment of problems which are common to all the utility industries (the valuation of the fixed asset base, the determination of allowable rates of return on electricity grids, pipeline systems and other networks, and the accounting principles for providing for the replacement of fixed assets). But such a commission might further reinforce the power which has increasingly been concentrated in the hands of central government in Britain. Nor is it clear that each commissioner could stay sufficiently in touch with the situation, problems and prospects of each utility industry to take appropriate decisions. The case for such a commission is weak as long as the regulation requires a lot of knowledge which is specific to the particular industries. The case would be much stronger if competition necessitates regulation of only the network infrastructures (in electricity, transmission and distribution) since these require common approaches and depend less on detailed knowledge of the industries. The question of undue concentration of power would remain, however.

A Combined Commission for Electricity and Gas

Nearly all the RECs now sell gas as a result of competition in the industrial gas market, and some or all of them may well enter the domestic gas market as it is progressively opened to competition. Similarly, gas suppliers may in future also sell power from CCGTs to electricity suppliers and/or electricity to final consumers. For present electricity or gas suppliers, selling both fuels would enable them to exploit their comparative advantage over new entrants of a comprehensive base of domestic consumers. This gives economies in meter reading and marketing, and therefore opportunities for price discounting based on lower costs in relation to new entrants who would probably have to incur high marketing costs for a time in order to attract domestic customers away from the long-established suppliers. A combined regulatory commission for electricity and gas is more likely than the present one-industry regulators to deal effectively with unfair competition by those who supply both fuels.

The position in early 1996 is that the Government regards the 'one-person' regulators as preferable to commissions but it recognizes that there may be a stronger logic for having a joint electricity and gas regulator after 1998 if problems in the domestic energy market warrant it.[2] Under the present policy, the main task of the electricity and gas regulators is to prepare for full competition in the domestic market from April 1998. If competition then worked so well that there was no need for regulation of generation or 'supply', the regulation of the monopoly infrastructures could be dealt with by a single regulatory commission covering all the utility industries. In early 1996, a 'wait and see' policy seems to be the best policy for the time being.

PUBLIC ACCOUNTABILITY

In Britain, public accountability at the highest level is normally exercised through Parliament – increasingly through Select Committees covering each Government Department, backed up with information and advice from various special interest groups including consumer, social and environmental groups. Regular Parliamentary scrutiny of the economic and financial performance of all the utility industries ceased at privatization. However, like all holders of public office in Britain, the economic regulators have been required regularly to give an account of themselves through annual reports, appearance before Select Committees and in other ways too. Also, some Select Committees – notably that on Trade and Industry with regard to electricity and gas – have published constructive and hard-hitting reports on utility industry issues. Therefore it cannot be argued that there is no Parliamentary scrutiny, or that the scrutiny which does take place is poor.

The case for a new Select Committee – call it the Utilities Committee – is that Parliamentary scrutiny of the utility industries needs to be broader, more systematic and less ad hoc than it has been since these industries were privatized. There is every reason for the existing Committees to continue their problem-oriented enquiries which involve economic regulation to a greater or lesser extent. The role of the Utilities Committee would be different. It would be required to assess: the effects of economic regulation in the various utility industries in relation to the criteria of 'good' regulation (to ascertain the extent to which the objectives are being met); the problems of monopoly regulation, including methods of valuation of fixed assets, determining normal rates of return, appropriate provisions for the future replacement of the network, and technical auditing of capital programmes; the regulators' monitoring of all aspects of supply licences in competitive markets; and the problems posed by takeovers and mergers, including the effectiveness of undertakings or other means of preventing unjustified transfers of funds out of the regulated business to the parent company or other subsidiaries, or the withholding of information from the regulators. This would provide not only the necessary overview of how the whole system of economic regulation is working, but also comparisons between the regulation of the different utility industries, and the necessary context for the more problem-oriented enquiries of other Select Committees. These suggested terms of reference are no more onerous than those of other Select Committees which have already established a high reputation.

If the case for putting a sharper focus on Parliamentary scrutiny of the utility industries were accepted, three practical issues would arise. Firstly, how could effective coordination of subject matter be achieved between the Utilities Committee and the other Select Committees involved? Coordination should not be a problem as there are effective procedures already to ensure coordination of subject matter between House of Commons Committees. Secondly, could Parliamentary scrutiny be exercised responsibly and without influencing the regulators while they are making their decisions and thus bringing political pressure to bear on the regulatory process? There can be no guarantee on this score, but past and present Select Committees have demonstrated repeatedly in parallel fields that it is possible for politicians to comment responsibly on a wide variety of public issues, including those involving sensitive material of a commercial or security nature. Thirdly, how could a new Select Committee be established? Formally this would be a matter for the House of Commons to decide, but in practice it is likely to be decided by the Government applying the Party 'whips' (or just influence) to block or support the proposal, which – through the Party system – would fail without Government consent.

During the first five years of electricity privatization, the Government showed no interest in the idea of a Utilities Select Committee, but in early 1996 it was reportedly considering the creation of an 'all-party parliamentary select committee to oversee utility regulation', as one of several possible measures to restore public confidence in the system of regulation before the next general election (which is due before May 1997).[3]

Consumer Councils with statutory backing and public funding were also part of the British approach to public accountability during the period of public ownership. Since privatization, consumer representation has generally been through consumer committees which report to the industry regulator. This is the case in the electricity industry where the consumer committees have rarely expressed a critical view of the regulation in the media. The consumer movement as a whole has been pressing for independent consumer councils for each privatized utility industry on the lines of the Gas Consumers' Council which, by a quirk of the legislative process, avoided abolition at gas privatization and ever since has provided an outspoken consumer voice, independent of and not frightened to disagree with the gas regulator. Independent consumer councils can take up complaints, press for higher standards of service, safety and appliance labelling, and press the consumer interest on pricing matters to help counterbalance the influence of those representing shareholder and company interests.

REGULATORY EXPLANATIONS AND EFFECTIVE CHALLENGE

The British economic regulators publish a great deal, including annual and special reports, consultation documents, speeches and press releases. But they are not required to, and often do not, publish a full explanation of their regulatory decisions. This has been criticized on two grounds: firstly, as being unacceptable in a democracy where the legislature and the courts are open to the public and the reasons for the decisions arising are stated and publicly available; and, secondly, because it reduces the effectiveness of judicial review as an appeal or arbitration procedure. If the grounds for a decision are not known, it can sometimes be difficult to decide whether the decision is illegal. For both reasons, public accountability and effective challenge, full explanations should be provided.

An alternative procedure for challenging a regulator's decision would be for the regulated company affected by it to request a referral to the MMC. But, as argued in earlier chapters, this is often daunting for the company because of the long time an MMC investigation takes and the possibly serious repercussions on the company if the MMC chooses to open up a range of further issues during the enquiry. It has therefore been suggested that either the MMC should find a 'fast-track' procedure, or a more effective appeal or arbitration procedure should be devised to circumvent both the MMC and judicial review.

The key issue is how to reconcile the need for effective challenge with the need for an effective regulatory process and avoid the danger of excessive legalism grinding the whole process to a halt. The Government currently admits to no problem and emphasizes that the regulators consult widely before taking decisions,[4] but this is no substitute for proper explanation of how the decision was reached and how different considerations were weighed in the balance.

Due to the unpredictability of the regulatory regime and the limitations of the appeals procedure, it has been argued that all aspects of the essentially five-yearly regulatory settlements or agreements should be incorporated into formal contracts.[5] The contract would include the new price control formula, how long it would run, precisely what it covers and does not cover, the background assumptions (cost of capital, asset base, operating expenditure and capital expenditure) and the mechanism to re-open negotiations if needs arise. This would help to provide the predictability and consistency which companies and shareholders want. But from time to time the public interest may well require reassessment of the price control in mid-term. This is because circumstances, information and the understanding of problems are liable to change and because of the built-in incentive for the companies to try to secure lenient determinations from the regulator. Mistrust of how monopolists can abuse their market power is a strong reason for scepticism towards the idea of formal regulatory contracts.

THE ROLE OF GOVERNMENT

Part of the case for privatization was that it would end political interference in the affairs of the utility industries and that the regulation would be independent and last only until monopoly had given way to competition. This was not to be. Not only will the natural monopoly elements of transmission and distribution continue to be immune to competition, but the Government determines the broad policy of competition and the general thrust of the regulation, appoints the industry regulators, makes referrals to the MMC, calls for reports from the Director General of the Office of Fair Trading and has powers to intervene in emergency. Although each determination by the regulator is independent, the Government has a major role behind the scenes and regulators cannot afford to ignore its wishes. That role is likely to be less problematic as experience of regulation accumulates, as the influence of the lenient initial price caps set by the Government at privatization is reduced, if the growth of competition reduces the use of the regulator's discretionary powers, and if the major political parties agree broadly on privatization, the structure of the industry and the general method of regulation. It could be a very different situation if the major political parties were in unbridgeable disagreement over the whole policy, or if a deep rift opened up between the regulator and the Government on the implementation of the policy. Whatever the degree of political controversy, the role of Government is inescapable. We examine three aspects.

Appointment of Regulators

Regulators are appointed directly by the relevant Minister – the Secretary of State for Trade and Industry in the case of the electricity and gas regulators. Partly as a result of this, the regulators tend to be seen as being closely aligned with Government policy on competition and market forces. Serious difficulties would arise under the British political system if regulatory decisions were far out of line with Government policy for any length of time. Difficulties for the regulatory system also might arise if the public and the opposition parties regarded the regulators as being too close to Government. The result would be a decline in confidence in the system and perhaps a change of regulator with each change of Government. An alternative method of appointment would be the more traditional method for senior public posts of open competition via the Civil Service Commission.

That would loosen the personal bonds of the regulator to his or her political master, both for the initial appointment and for subsequent renewals, and end the suspicion that the regulators are appointed because they share the same ideological views as the Government. However, the selection panels would still need to ensure that candidates' experience and views make them suitable to do the job within the broad guidelines of Government policy.

Government Policy

It is widely understood that the regulation has to take place within a framework of Government policy and that the framework will inevitably evolve through time. But if there were thought to be too much 'back-seat driving' from Whitehall, public confidence in the regulation would tend to decline. 'Back-seat driving' has been more of a problem so far for gas than for electricity. This was largely because the Government tried to create a competitive structure and published a timetable for introducing competition in electricity supply, whereas in gas it simply transferred a vertically integrated, national monopoly into the private sector. When competition in gas did not occur as a result of the change in ownership, the Government expected the regulator to use his powers to promote competition and thus change the structure indirectly. That was part of the reason for the different styles of regulation – the highly adversarial, David and Goliath style in gas from 1989 to 1994 compared with the more relaxed style of the electricity regulator who seemed confident that full supply competition from April 1998 would lead to competition in generation and thereby secure significant price reductions for final consumers – especially domestic consumers. The decision in December 1993 to introduce competition for domestic gas consumers in 1998 necessitated many consequential decisions involving close collaboration between the DTI and OFGAS. The rapid incremental change in gas policy and regulation since 1990, and the controversies and uncertainties which this involved, underline the need for Government to try to avoid adopting an industry structure at the outset which is inimical to competition if competition is the objective and, where a major change of direction occurs, to publish new Government objectives rather than just to rely on regulators to use their discretionary powers to bring about the change indirectly. In both electricity and gas it is most likely on the evidence so far that Government will continue to have a major role alongside the regulator, and sometimes this will lead to a difficult relationship.

Wider Economic and Social Issues

A major criticism of the economic regulation in Britain has been the narrow focus on competition and internal costs and the neglect of wider economic and social costs and benefits. A major example is that it has proved difficult to fund energy-saving schemes through levies on gas and electricity bills or by adjustments to the RPI–X price control. After the Government decided to use the Energy Saving Trust as part of its policies in connection with the 1992 Rio Conference on environment and development it emerged that it was inappropriate as well as of doubtful legality for the Government to require the gas and electricity regulators to impose substantial levies on consumers in order to finance the large expenditures proposed by the Trust to promote energy efficiency. Another major example is that large-scale redundancies in the privatized utility industries have been encouraged by RPI–X price regulation, while very large job losses in the coal industry resulted from the attempt to promote competition in electricity generation through the 'dash for gas'. These developments have increased total unemployment and the social security costs borne by the taxpayer. This is not meant to imply that no redundancies would have occurred without privatization, although they would probably not have been on the same scale. Like other industries, the utility industries have been heavily affected by the revolution in information technology and other technical change, and these industries have shared in the trend in Britain towards the contractualization and part-time nature of work. But if wider social and economic issues are to be dealt with properly, the rules for doing so should be decided by the Government and Parliament rather than by appointed regulators who do not have the political authority and in some cases (eg for energy efficiency through the Energy Saving Trust) appear not to have appropriate statutory powers to impose the necessary levy on consumers' bills. It is up to Governments to take responsibility on all such 'externality' issues.

REGULATING THE NATURAL MONOPOLY NETWORK

THE METHOD OF REGULATION

The natural monopoly elements of transmission and distribution account for around 30 per cent of the average cost of electricity (excluding the fossil fuel levy). This represents the second largest proportion of electricity costs after generation, whereas the cost of 'supply' (or retailing) is much smaller. The method adopted from the outset for regulating the natural monopoly elements (as well as for the 'supply' business until it is fully exposed to competition) is price regulation, which was considered to be superior to rate of return or profit regulation, particularly in stimulating internal cost efficiency.

But there are three reasons why the distinction between price and profit regulation is by no means clear-cut. Firstly, the introduction of competition in electricity and gas necessitated non-discriminatory charging schemes to allow competitors access to the transmission and distribution systems. This involved estimating the value of the relevant capital assets and deciding appropriate rates of return for existing assets and for new investment. To this extent the British system of economic regulation is already a hybrid which combines elements of both price and rate-of-return regulation, where the latter is designed to produce a 'normal' rate of return on the fixed assets of the network. Secondly, because the price controls are usually set for five years, the incentive for cost-efficiency can be blunted by the incentive for the company to retain inefficiency in the latter part of the period before the price control is re-set (eg by postponing cost-saving investment in the hope of obtaining a relatively lenient price control), thereby enabling the company to reap higher profits for a further five years. This is more likely to occur where the regulator was too lenient at the previous price-control review. Thirdly, in the long-run, price regulation and rate-of-return regulation will tend to produce broadly similar economic results, at least in terms of profits if not also prices. To retain the necessary social and political acceptability, price regulation would need to produce satisfactory levels of profit, while rate-of-return regulation would need to produce acceptable levels of price. The reason is that the great majority of electricity, gas, water and telecommunications consumers are also voters whose criticisms and complaints, if sufficiently loud and prolonged, represent political risk for the Government. In the event, the outcome as far as electricity is concerned has been far from satisfactory, as shown by continued press criticism of the levels of profits (and boardroom pay and perks which, although not strictly due to regulation, have been a focus of public criticism). The RECs have come in for special criticism because a very high proportion of their operating profits have derived from the monopoly distribution business (87 per cent in 1994/95) and have represented a margin of no less than 42 per cent on turnover. Likewise, the National Grid Company has had a high profit margin in relation to turnover (46 per cent in 1993/94).

As in the other privatized industries, the price controls in the electricity supply industry were set by the Government for the first three or five years (depending on the activity). They proved to be remarkably lenient and therefore attractive to shareholders. Other things being equal, the changes introduced under the 1994/95 distribution price review (discussed in Chapter 5) should reduce both the RECs' profits and the level of criticism over the next five years. But is there a better method of regulation which would divide the economic rent more fairly between shareholders and consumers, and not make domestic consumers wait so long to obtain their share of the benefits?

The RPI-X price control is designed to promote economic efficiency and the normally five-year interval between re-setting the price control reduces regulatory uncertainty for the shareholders and the companies concerned. The system depends on choosing the level of X which anticipates the rate of cost savings which will be achieved over the following five years. If the actual cost savings turn out to be above the value set for X, the result is additional operating profits for the company (and probably extra dividends for the shareholders) for up to five years. Conversely, if the value set for X exceeds the level of future cost savings, profits would fall and the level of investment and quality of service would suffer in the longer term. If profits were widely considered to be too high or if there was under-investment and poor service quality, growing criticism would probably be aimed at Government – just as it was in the days of public ownership. This is because Government was responsible both for privatization and for the system of regulation.

It is difficult even for the companies involved to predict cost levels and trends accurately over a period of five years. It is doubly difficult for the regulator since he or she is largely dependent on the regulated companies for cost information and performance indicators. In order to get a lenient price control, the companies have an incentive to exaggerate their cost levels, underestimate the scope for future efficiency gains, and overestimate their future capital expenditure programmes. The scope for effective yardstick competition using detailed comparisons among the various RECs and with overseas electric utilities is more limited than the Government thought at privatization. As was known under public ownership, the geographic, economic and other differences between the various RECs result in wide differences in both performance and costs which make it difficult to draw useful lessons. The companies involved are nevertheless still interested in 'benchmarking' as a means of comparing costs and reducing jobs.

In re-setting the price controls, there is much scope for judgement – especially regarding the 'normal' rate of return on the asset base on which the rate of return is measured, and in assessing the future trend of controllable costs (those costs which are supposedly within the control of the company's management). The difficulty of carrying out the five-yearly re-setting of the price control without significant error is indicated by the various stages involved in the 1994/95 distribution price control review. The regulator was assisted at each stage by consultants and the RECs were asked for detailed business plans and projected revenue and capital expenditures. Comparisons were made of companies' costs in the previous year to cast light on the factors which influence costs, the effect of different circumstances and any underlying differences in efficiency. A view was taken on the appropriate value of the assets of the distribution business, taking account of what shareholders paid for the assets at privatization as opposed to their current book value. The cost of capital was estimated using various sources. Finally, all the information on operating costs and capital costs was translated into sums of revenue over the following five-year price-control period. However, as the regulator admitted, estimates and calculations 'can only be a guide informing a judgement on the appropriate level of allowed revenue'.[6]

The risk of excess or inadequate profits could be reduced by switching to rate-of-return regulation. But if the possible effects of this in terms of operating efficiency and 'gold-plating' of investment[7] were considered unacceptable, there would be two main options. Firstly, the risk of too high or too low profits could be reduced by shortening the period between re-setting the price controls from five to, say, three years. Shortening the period would limit the political risk, but it would not be liked by the companies and their shareholders since it would increase regulatory uncertainty and might reduce the incentives for both investment and efficiency. The second option would be to retain RPI-X but to attach a 'sliding-scale' mechanism to it, so that above a certain level, the economic rent would be divided automatically between shareholders and consumers, for the latter as price reductions. This would not make the problem of choosing the value of X any easier, but it would greatly reduce the problem of excess profits since these would be shared with consumers. This method would retain both an incentive for efficiency and the stability provided by five-yearly price controls. The most controversial aspect seems likely to be deciding the proportions in which any surplus above the 'normal' rate of return is to be divided between consumers and shareholders.

CAPITAL EXPENDITURE AND RATES OF RETURN

The capital side of the regulation tends to be even more difficult than the revenue side. The regulation of the transmission and distribution systems has to allow the companies to derive an appropriate rate of return on the fixed assets purchased at privatization and on new investment since then. It requires taking a view on the appropriate value of the fixed assets allowing for the fact that the book value (based on current cost accounting) is likely to be far higher than the heavily discounted sum which shareholders have paid for the assets in the various privatizations, including electricity supply. The regulation also has to allow for the future capital expenditure necessary to maintain and modernize the network infrastructure, including the capital expenditure likely to be incurred in the next five-year price-control period and provisions for assets which will be replaced long after the next price-control period.

There is an obvious incentive for the companies to over-estimate future capital programmes before a price control review in order to persuade the regulator to allow large sums for this purpose, and to use the resulting capital underspend to boost operating profits. In this situation the simplest solution might be for the regulator to intervene as soon as the discrepancy has come to light and to tighten the price control formula accordingly (by increasing the value of minus X). However, the reasons for capital underspends need to be analysed and understood, for example, the extent to which they were due to greater cost-effectiveness, technological advance, or a reduction in the work previously considered necessary.

To establish tighter control over capital programmes, the regulator would need to carry out detailed technical audits of the scope, content, materials prices and wage costs of the whole capital programme. These would undoubtedly further increase the spread and the cost of regulation. In any case the criteria for 'adequate capital programmes' are difficult to establish, particularly for transmission and distribution systems which have long replacement cycles (eg 40–60 years) and where the scope for and effects of technical change – such as superconductivity and small-scale, decentralized generation embedded in the low voltage distribution system and thus not using high voltage transmission – are impossible to predict at this stage.

Decisions about the replacement of long-lived capital assets, especially those of the size of electricity and gas networks, have always involved major judgements about both the appropriate period and method of depreciation. These involve the question of how much of the cost of the eventual replacement of the system should be borne by present customers through current prices which cover the estimated full replacement cost of the assets, as opposed to reliance on future borrowing in effect to pass some or all of the cost burden on to the generation which will benefit from the use of the new assets. The long replacement cycles of electricity and gas networks, combined with the fact that very large investment in these networks was made in the period of public ownership and that major replacement programmes are not expected to be necessary over the next 15–20 years, make these network systems large sources of cash surplus which yield a substantial annual interest income for the companies and their shareholders.

In these circumstances there is an obvious temptation for the electricity and gas regulators to press for a 'cash flow' approach to achieve a somewhat closer match between the level of capital expenditure expected in the next price-control period and the depreciation charges to be allowed in that period, thus reducing the cash surplus. This would be a big departure from the initial approach of equal annual depreciation provisions to ensure that the accumulated sum eventually matches the estimated full cost of replacing the network. Doubtless the companies would object on grounds of financial prudence and because the cash surplus, and therefore the future flow of interest income, would be reduced.

The extreme version of this approach, known as 'Pay as you go', would allow the companies to charge in the accounts only the capital expenditure expected to be incurred in the next price-control period as opposed to depreciation provisions based on the full replacement costs. As that could have quite dramatic effects, some people are calling for a compromise which would reduce but not eliminate the cash surplus arising from the replacement provisions and which would not throw the whole burden of financing system replacement onto a future generation. Given that full competition in both electricity and gas supply is due to start in 1998, the regulator has an obvious temptation to deliver lower prices in order to justify the claim that competition benefits all consumers. Reduced transmission and distribution costs would at least reduce the problem of 'losers' in the competitive domestic market. This issue appears relevant to all the utility industries in view of their network nature, and it may need a generic determination by the MMC.

OTHER REGULATORY ISSUES

OWNERSHIP ISSUES

Taken together, the privatization legislation, the prospectus, the licence and the price controls represented a kind of social contract embracing the companies, shareholders and consumers. At the same time, Government 'golden shares' in the RECs safeguarded the structure adopted at privatization for the first five years. During those five years, however, it was widely expected – because of the high profits of the RECs' distribution activity and the opening up of the remainder of the 'supply' market in 1998 – that there would be takeovers and mergers if the 'golden shares' were allowed to lapse at their expiry date in March 1995. That proved to be correct. Although the distribution price control was tightened considerably in early 1995/96 (in response to Northern Electric's access to large cash reserves to fight the Trafalgar House bid), that action by the regulator did not stem the tide of takeover bids. The DTI was consistently lax about the bids, overruling the OFT and OFFER on the issue of referrals to the MMC and only belatedly agreeing to refer cases involving vertical re-integration.

Chapter 10 sets out the takeover activity involving the RECs until the end of 1995 and the competition policy issues which have been raised, particularly the extent to which vertical re-integration should be allowed. Actual and proposed changes in ownership of the kind which occurred in 1995 also raise issues relating to the effectiveness of the regulation of the monopoly distribution activities which are a source of most of the RECs' profits. Given the diversity of the takeover activity – with RECs joined with US utilities, conglomerates, UK privatized water companies, and with a Scottish generator, OFFER's ability to apply meaningful yardstick competition is even more limited than before. This is true even if separate regulatory accounts continue to be required for the RECs' distribution businesses. Moreover, the absence of separate published accounts for electricity undertakings which have been taken over reduce the public accountability of the privatized ESI.

Information on the cost of capital, which is important to OFFER in deciding whether price regulation is working satisfactorily, also becomes more difficult, since such information is normally derived from the stock-market performance of companies rather than the performance of their separate activities whether regulated or not. Above all, the takeover activity has increased the inherent problem of the imbalance in the information available to the regulator compared with the usually much greater amount of relevant cost information of the company owning the regulated business. Identifying and ring-fencing the true costs of the regulated distribution activity is much more difficult to achieve given the ability of large, complex companies to re-allocate a proportion of their costs and cash flow between their various activities in order to maximize profitability or secure tax advantages. Two possible approaches to reduce the disadvantage of the regulator are: firstly, to stipulate before a takeover is allowed that the parent company must provide information requested by the British regulator about their business activities (not just the regulated activity in Britain); and secondly, for the British regulator to seek to establish cooperative relations and information exchange with the relevant foreign regulators, especially in the USA (in much the same way as applies in international tax matters).

A complex problem for both OFFER and OFGAS would arise if the RECs and/or their parent companies were to enter the public gas supply market on a long-term basis, or likewise if gas suppliers were to enter the public electricity supply market. By early 1996, most of the RECs were selling gas in the industrial market. Whether they would continue to do so given the large fall in profit margins there and whether they would launch into domestic-market gas competition which has different characteristics and requirements, is not yet (in May 1996) clear. But if the joint selling of electricity and gas occurs on an appreciable scale, there will be a strong case for a joint electricity and gas regulator.

The main attraction of RECs to potential predators has been the substantial positive cash flow from the monopoly distribution activities, net of investment requirements. The extent to which such surplus cash is diverted to other activities owned by the new parent companies could well become an issue in future. It is clearly bound up with the question of the monitoring of capital expenditure programmes for the 'wires' infrastructure, as discussed above.

There is here a link with the question of the public acceptability of the regulatory regime for electric and other utility industries which could be jeopardized if the profits from the natural monopoly parts of the business (in electricity, especially the 'wires' business of the RECs) are seen to be going overseas or to unrelated activities of parent companies. Nothing is likely to be more corrosive of the 'social contract' which underlies the system of economic regulation of the utility industries than if the public service ethos is widely seen as having been replaced by the profit motive, if financial surpluses from the UK electricity activity are used for the benefit of parent companies' overseas businesses or entirely unrelated businesses;

or if consumers feel they are being exploited by multi-utility monopolies. This emphasizes the danger of the view that the services provided by the utility industries are like any other commodities while forgetting that the public needs universal public service provision of a high standard of reliability and safety and at reasonable and affordable prices.

REGULATION TO PROMOTE COMPETITION

Where competition is the objective but the privatized industry structure does not encourage it and Ministers do not want to revise the privatization legislation, competition can only be secured through regulatory intervention and pressure. The main case in Britain has been the re-structuring of the industrial gas market from 1989 to 1994, relying heavily on the regulator's use of various artificial props to assist new entrants and restrict the incumbent (British Gas) to a diminishing and less profitable market share.[8]

The main case so far in the electricity industry has been the requirement placed on the two main generators to divest 6 GW of their fossil fuel plant (see Chapters 4 and 10). It is still unclear (in early 1996) how far the divestment of the 6 GW of plant will reduce the duopoly power of the two main generators. The point at issue concerns how the divested plant will actually be operated. To restrict the duopoly power requires the divested plants to operate intermittently (on mid-merit) where they would remain in competition with the fossil fuel plant owned by the two main generators, and where they could affect the Pool price. That is the rationale behind the divestment. But the new owners might wish to obtain base-load contracts for those plants since, even at a lower price per kWh, base-load contracts would probably increase their total profits. If the divested plant were to operate on base load, they could not affect the Pool price and the duopoly power would remain.

As far as competition in 'supply' is concerned, the implicit assumption behind policy has been that the ending of the franchise of the under-100 kW market (overwhelmingly the domestic market) in 1998 will stimulate lively competition among the RECs and between them and new entrants. Prediction is difficult under such uncertainty but several factors suggest that competition may be limited.

Profit margins on the RECs' 'supply' business are small compared with those on their 'wires' business, and each REC will retain the profit on its wires business even if it increasingly loses its supply business to competitors. If there were a good deal of customer inertia against changing suppliers, the RECs would not need to reduce prices much (if at all) in order to retain a large customer base in their traditional local supply area. Retaining a large base of customers could be important in several respects: keeping the average costs of marketing and meter-reading as low as possible, providing a good customer database for establishing a profitable domestic gas supply business, and offsetting the cost of billing with income from direct mail for other companies. By contrast, competing suppliers would face the likelihood of high marketing costs to become established in new areas, and when established they would derive only relatively small profits from the supply business. In these circumstances, an REC might choose to defend its existing domestic supply market and if that eventually proved too costly, to let supply go, retain the profit on the wires business, and seek to diversify profitably.

The dynamic in this situation could come from two directions. One would be from RECs seeking to build up market share by 'cherry-picking' more profitable customers and hoping to build up ancillary sales of appliances, double glazing, cavity-wall insulation and so on. The other direction would be from new entrants. It is not clear who these would be, but it has been suggested that some of the big multiple-store groups could provide strong competition. But why they should enter a low-profit business, especially when competition is cutting profit margins in their core business, is unclear. Not only that, the RECs have their profits from their 'wires' business which, unless there is strict monitoring and inspection by OFFER of the RECs' methods of allocating joint overhead costs, the RECs (or their parent companies) might find ways of using in order to cross-subsidize their 'supply' business.

In these circumstances the Government or OFFER could initiate anti-monopoly action aimed at making the RECs divest their wires business and leave them with the relatively low profit margins of the supply business. This would be required only if the cross-subsidy could not be prevented through regulation. Alternatively, the RECs might be required to divest their supply business; but this would be possible only if it were known that numerous competitors wanted to enter the supply business. The latter would be more likely if the profits on supply were higher than they have been so far.

Conclusion

Four things have become clear about the British system of economic regulation of the privatized utility industries. Firstly, as natural monopolies, transmission and distribution will have to be regulated indefinitely and the method of regulating them is already a hybrid of price regulation and rate-of-return regulation; and it may hybridize further if sliding-scale regulation is added. Secondly, there are major difficulties with the method of regulation: one is in trying to set X in the price control formula at the level which will correctly anticipate the cost savings over the next five-year control period; another is in dealing with the problems on the capital side of the regulation. Thirdly, there are strong cases both for changing from the present 'one-person' regulators to regulatory commissions and for merging the electricity and gas regulators if it becomes clear that significant amounts of electricity and gas are being sold into the same markets by the same suppliers. Fourthly, there will be a much stronger case for merging all the regulation into a single regulatory commission covering all the utility industries if competition works so well that it is necessary to regulate only the natural monopoly infrastructures.

Although regulation can affect incentives and therefore the efficiency of the utility industries, primarily it serves as a 'social contract' which seeks to balance the interests of consumers and shareholders – chiefly in distributing the economic rent in a way which retains the broad acceptability of the various stakeholders, especially consumers and shareholders. A high level of dissatisfaction among shareholders indicates a risk of declining confidence in the company, an exodus of capital, and under-investment. The public acceptability of the regulation has been low because of consistently high profits and boardroom pay increases, which emphasize the profit motive and demote the public service ethos.

Much of the criticism of the economic regulation of the privatized electricity industry over the past five years has stemmed from the Government's lenient settlement at privatization which resulted in large profits for low-risk monopoly activities and long-delayed benefits for the great bulk of consumers. This source of dissatisfaction may now diminish sharply due to changes in the regulation in 1995. The public acceptability of the economic regulation will increasingly depend on the success or otherwise of the policy of introducing competition into the domestic market, including the extent to which competition in 'supply' is effective in reaching all domestic consumers, whether it stimulates competition in generation, and whether it is compatible with maintaining supply security for the individual consumer and for the electricity system as a whole. The latter and other matters of a strategic nature are taken up in the next chapter.

12

Strategic Government and Corporate Issues

Steve Thomas

Introduction

Over the hundred years of its existence, almost every aspect of decision-making of the electricity supply industry (ESI) worldwide has come under the direct or indirect influence of national Governments. The declared rationale for the exertion of Government influence has consistently cited the need to ensure that the ESI's decisions fitted in with strategic national objectives, in short, to secure the national interest. In some cases the aim of Government intervention has been to guarantee that the supply of an increasingly vital public service was cheap, reliable and universally available. In others, it has been a result of the sheer size of the electricity sector which meant that decisions taken within the industry often had significant repercussions for other sectors of the economy which are dependent on it, such as the fuel- and equipment-supply industries.

There are additional practical reasons why electricity supply industries are used by Governments to achieve objectives other than those of the industry. Electricity supply industries generally operate only within a single country and do not have interests in other sectors and the companies are therefore, in some senses, easy targets. The monopoly status that the industry has enjoyed also means that the cost of such interventions can often be passed on to consumers with little risk to the profitability of the companies.

The strategic dimension sometimes arises because, more than most industries, electricity supply tends to impose burdens or confer benefits on society which are not reflected in the costs that fall directly on the industry. Where such large burdens or benefits arise, for example through environmental impacts, it is argued that Government has a responsibility to set a policy framework to ensure their costs are transmitted to the industry causing the impact. As a result, over the past two decades, the need to control the environmental impact of the activities of the ESI has become an increasingly important obligation on Government.

This chapter discusses the effects that restructuring and privatization have had on the way Government interacts with the ESI. It identifies five main areas of historic Government involvement: system planning; utility purchasing policy; technology choice and development; national macro-economic and social considerations; and environmental protection. In each of these areas, privatization changed, and in most cases diminished, the role of state. A sixth area, promotion of energy efficiency, in which many argue Government should have played a more active role before privatization is also discussed.

These areas of influence are examined in three parts: first, the factors are identified which, in the past, have motivated Governments to require strategic national considerations to be included in the commercial decision-making of the ESI; second, there is a discussion of why pressures have arisen to change the way in which Government interacts with the ESI; and third, there is an examination, in the light of the experience since privatization, of the advantages and the risks to the national interest that relinquishing these traditional avenues of influence will have. Finally, the development of the future structure of the electricity supply industry is explored. The component parts of the industry are being integrated, by takeovers and mergers, more fully into the wider international, industrial corporate structure. These changes mean that electricity supply industry companies will no longer be contained within national boundaries and will increasingly have interests in other industrial sectors. This will make Government influence more difficult to exert.

There is no clear dividing line in Government interactions with a nationalized industry between strategic interventions and the legitimate management oversight of a company that the British Government, as owner of the industry, was statutorily required to fulfil under the 1945 Ministry of Fuel and Power Act. However, in this chapter we concentrate on the former, that is areas where consideration of the broader national interest was the main objective, rather than normal shareholder influence.

STRATEGIC ASPECTS OF ELECTRICITY SUPPLY

From 1947 in the UK, Government was able to ensure that strategic considerations were taken into account directly through its ownership of the industry. Pressure to reduce Government involvement in ESI decision-making arose partly due to doubts about the effectiveness of the measures adopted in achieving their aims. As a result of this pressure, one of the stated objectives of electricity privatization was to reduce greatly the scope for Government to impose strategic considerations on decision-making. This was characterized by advocates of the reforms as 'Government interference', which was seen as having adversely affected the economic efficiency of the ESI.

Pressure has also arisen because of changes in market circumstances which seem to have reduced the risk of the market failures that Government involvement was meant to guard against. A common thread amongst the criticisms of state intervention has been a perception that this led to the loss of some of the advantages of market discipline and that, as the risk of market failures diminished, the scale of the compensating benefits was decreasing. Preserving or nurturing certain favoured resources for strategic reasons often meant they were insulated from competitive pressures and this led to inefficiency and inflexibility. For example, the British coal industry was protected by the Central Electricity Generating Board's (CEGB) procurement policies even though the reliability of world fuel markets had increased, particularly for coal, as the size of the market had grown and prices stabilized. There was also increasing confidence that the oil market, which tended to drive the prices of other fuels, was much less at risk from politically inspired disruptions. International fuel prices were low and serious disruptions in international fuel markets were no longer feared.

In addition, Government decision-making was seen to be inevitably too remote from the market for it to be sufficiently informed. Since 1979, a strong theme of the ruling Conservative Party Governments' policy has therefore been to reduce Government interference in industry in general. As a result, the Government has frequently been criticized for having no energy policy, although this has been disputed by Government Ministers. To some extent, this dispute is one of semantics. The Government clearly cannot avoid having policies which affect the energy sector and it has general policies, such as privatization, which it would be inconsistent not to apply to energy. The more significant point is whether Government really has surrendered its traditional role in energy policy which was based on a view that energy is of such strategic importance that Government had a special role that it did not need to fulfil in other sectors of the economy. Privatization did lead to a loss of many of the traditional lines of influence, but Government influence is still strong, especially through economic and environmental regulation.

Historically, Government influence on the electricity supply industry can be identified in five main areas: system planning; utility purchasing policy; technology choice and development; national macro-economic and social considerations; and environmental protection. The development of policy in each of these areas and in promoting energy efficiency is now considered.

SYSTEM PLANNING

The Old System

Under the previous regime, the nationalized companies had a duty to ensure that all reasonably foreseeable demands for electricity were met. In the new system, it is implicitly assumed that competition in generation and supply will mean that the planning processes that existed will no longer be needed and that market forces alone will ensure supply security. The main issue examined in this section is whether the new British electricity system can deliver a similar level of supply security as the old arrangements.

Historically, a constant concern for Government was the need to ensure that electricity supply was cheap, reliable and universally available. With the increasingly integral role of electricity in the economic and social fabric of society, came a growing perception of a need to oversee the planning of the supply industry. This was particularly so in the postwar period of reconstruction, full employment and economic growth. The consequences of a failure to supply electricity reliably and at prices which did not disadvantage industry or put at risk the welfare of the population were seen as too high to tolerate weaknesses in the electricity supply chain.

A tacit 'regulatory bargain' emerged, in Britain as elsewhere, between Government and the generation companies which had a franchise area within which substantial monopoly rights were granted. In return for monopoly rights, a utility would have a responsibility to ensure security of supply. This inevitably required forecasts of maximum demand, since this determines the amount of capacity required, and forecasts of the capacity that would be available to meet it. As a result, planning procedures were established in the ESI which required expertise in forecasting short- and medium-term demand patterns, the impact of unusual weather patterns and the availability of generating capacity, particularly at the time of peak demand.

After World War II, Britain faced shortages of generating capacity and fuel, and Government did not have confidence that privately owned companies or companies owned by local authorities could be relied on to solve these shortages. In part these doubts arose because of the huge pent-up, competing demands for investment capital that Britain faced from the utility industries, other basic industries and housing in the immediate postwar years. Nationalization of the British ESI in 1947 was expected to ease this problem by allowing Government automatically to take a central role in the strategic decision-making of the industry. Electricity nationalization was implicitly based on a belief that central authorities, accountable to Government, were required to plan a national electricity network. For the long-term, it would also ensure that, once the immediate shortages of coal and power-station equipment had been overcome, all foreseeable electricity demands would be met.

Pressures for Change

From then, until the time of privatization, there were few doubts expressed about the basic need for electricity-supply system planning. However, the planning system within the UK electricity supply industry, like the ESIs in many other countries, had a poor reputation on a number of grounds. Demand forecasts had tended to under-estimate demand growth when it was high and over-estimate demand growth when it was falling. After the 1973/74 oil crisis, the demand forecasts and capacity planning procedures were widely seen as being influenced by the desire to justify a large nuclear power programme.

Government monitoring of these procedures appeared to have been ineffective in improving planning, and Government demand forecasts frequently suffered from the same weaknesses as utility forecasts. Some argued that the tendency of the ESI to over-forecast, after the severe capacity shortages of the early 1960s, was the inevitable result of a monopoly system where the costs of over-forecasting could be passed on, often with little scrutiny, to consumers. By contrast, the consequences of under-forecasting and the resulting reductions in service quality were public humiliation of utility management. Management therefore had a strong incentive to err on the side of over-investment.

By the time electricity privatization was proposed, electricity demand growth had, for many years, been well below the annual 7 per cent rate that had been the 'iron rule' for the electricity supply industries in developed countries since their inception. As a result, the focus of ESI corporate planning had already shifted from demand forecasting and new-capacity planning to achieving cost-minimization.

Advocates of the introduction of competition to generation through privatization argued that this would make it more likely that supply and demand were accurately matched because market pressure would ensure costs were minimized. Privatization would also make it more likely that if over-investment did occur, its cost would be borne mainly by shareholders rather than by consumers or taxpayers, as had always previously been the case; and this would tend to make the generators more cost-conscious than the nationalized industries had been.

An additional factor for some countries in mainland Europe, which brought into question the assumed need for national electricity system planning, was doubt about a tacit assumption in national electricity-supply system planning: the belief that all countries must plan to have sufficient generating capacity to supply all their own power needs. Countries such as Italy had become long-term power importers, albeit not by design, with no apparent detriment to system supply-security. However, for Britain, its island status and the high cost of sub-sea interconnectors meant that the flexible trading of a significant quantity of the nation's power in international markets was not a realistic option. The link between France and England can only supply about 5 per cent of Britain's electricity needs and the high fixed cost of such links means that an increase in the capacity of connections to mainland Europe would be hard to finance, especially now that the risk of building such a link would tend to fall on private investors, not electricity consumers.

Experience Post-Privatization

After privatization, the unavoidable retention of the monopolies for transmission and distribution meant that specific accountable bodies, the National Grid Company (NGC) and the Regional Electricity Companies (RECs), continued to have an obligation to ensure that the infrastructure was adequate to meet the demands placed on it. But for generation this was not the case: the implicit assumption behind the Power Pool for England and Wales was that market forces would ensure that sufficient generating plant was available. All generating plant must be owned by companies licensed by the regulator and the plant must be registered every six months with the Pool and the NGC. But plant owners cannot be forced to place a bid into the Pool on days when demand is high and the decision to retire old plant is entirely at their discretion and will be based on whether the plant is expected to make a profit. Likewise, the decision to build new plant is based on its expected profitability and, while new plant is subject to Government planning consent and requires the approval of the NGC, like plant retirements, the decision to proceed will take no account of whether new plant is needed to meet demand securely.

The hope is that any plant that makes a loss will not be required to maintain reliable electricity supplies and can be retired with no detrimental effect on supply security, and that the Pool will give strong enough price signals to the market to invest when new capacity is needed. The volume of plant ordered under the 'dash for gas' has been more than matched by retirements of coal-fired plants. However, a small plant surplus at the time of privatization, and favourable weather conditions since then mean that this has not led to any serious supply-security problems yet. The ability of the competitive generation market to provide supply security will be more fully tested when the strategic reasons for building new plant that prevailed in the 'dash for gas' no longer apply. It will also be tested when one or more of the rare events occur that electricity systems must be designed to withstand – severe weather conditions, or a large number of coincidental plant break-downs, for example.

Assessment

While the system has not failed since privatization, the six years' experience since then have done little to prove the ability of the market-based generation system to provide supply security. Low demand growth has meant that there was little need to build additional capacity. The large volume of combined-cycle gas turbine (CCGT) orders was motivated as much by strategic considerations as by projections of profit, and certainly had little to do with meeting a need for new capacity. The owners of this plant had little difficulty in obtaining consents for the sites chosen because of the small physical size of CCGTs, the absence of fuel stocking facilities and the low emissions. It is highly likely that any proposal to build a large coal-fired plant, particularly if it was sited outside coal-mining areas and did not use British coal, would have been vigorously opposed in the local planning procedures.

As argued in Chapter 4, the CCGT orders placed were either ordered under contractual terms which allowed little risk to the owners, or by the two large generators whose duopoly power meant they could absorb the risk. Thus, the belief has not yet been tested that market forces would ensure that supply and demand were efficiently matched because of the shift of the economic risk of building new plant from consumers to shareholders. Experience therefore provides little encouragement that market forces will provide the right signals, ensuring that plant is retired only when it is not needed for system security and that the right amount of new plant is ordered to meet demand growth.

If market forces alone are not sufficient to ensure supply security, it would seem necessary to re-introduce a greater element of system planning. However, it would be difficult to devise a system which did not compromise one of the basic tenets of a free market, that of free entry into and exit from the market. A planning process would require that the owners of existing plant would have to give a commitment of several years, long enough to build a replacement, for which their plant would remain in service. Generating companies would be unlikely to agree to these conditions unless there was some means of compensating owners of uneconomic plant for keeping them available.

It would also require that when a need for new plant was identified, some mechanism would have to exist to ensure that this capacity need was met. In a market-based system, this should be done by some form of competitive process. But the plant selected could only be financed if, in return for the owner of the plant sticking to the costs contained in their bid, the winner of the contest was given some long-term assurance on the utilization of the plant and the price paid for power. A universal Power Pool into which all plants must bid on a daily basis, which is the central idea behind the new British generation model, would be hard to sustain if the assurances provided by the planning procedures to owners of new plant totally removed a significant volume of plant from its influence.

UTILITY PURCHASING POLICY

A second aspect of ensuring the reliable supply of electricity concerned the purchases made by the industry in areas where the markets for these purchased commodities could not be relied upon, or simply did not exist. Utility purchases of fuel and large items of equipment were of particular concern in this respect. The sheer size of these purchases meant that their social and macro-economic implications were also important factors in determining policy. However, a constant theme amongst some critics of the old system was that the utilities were paying too much for fuel and equipment. The main issue examined in this section is whether it is likely to matter that ESI privatization has meant that there is now little room for Government to influence the fuel and equipment procurement policies of the ESI.

The gap between world market prices for coal and heavy electrical equipment was cited as evidence for the claim that Government interference was imposing heavy, unjustified extra costs on electricity consumers. The size of this apparent price gap almost certainly over-stated the scale of the problem. If Britain had bought all its power-station coal on the world market, the volume of coal traded would have increased dramatically, probably raising prices as a result. Similarly, international trade in equipment was only a small part of the market and so international prices could not have been assumed to hold if a much higher proportion of the output of the industry was traded. However, even if the scale of the price discrepancy was over-stated, as argued in Chapter 3, there are grounds to suggest that the CEGB did not place as much competitive pressure on its suppliers as it could and should have done.

It is in the area of utility purchasing policy that the relaxing of Government influence has been most conspicuous. The size of the British coal industry has been dramatically reduced and procurement of equipment has clearly been based purely on the merits of the bid rather than the nationality of the bidder.

Fuel Supply

The Old System

Worldwide, ESIs were generally founded on the use of indigenous energy resources, usually either hydro-power or solid fuels such as coal, lignite or peat. The general situation in most developed countries by the mid-1960s was that only small hydro-electric resources remained to be exploited and the cost of indigenous fuels was not competitive with oil available in rapidly increasing volumes from the Middle East. Utilities therefore began to look beyond their traditional indigenous options for new generating capacity with the cost of generation from imported heavy fuel oil as the bench-mark. Cost projections from the USA suggested that nuclear power was, by then, also a serious option. National primary energy supply strategies were developed to determine how electricity demand growth would be met, and the rate and extent to which local fuel industries were run down became increasingly important.

These pressures were also apparent in Britain. In the 1960s, coal supplied from British mines to British power stations was beginning to seem an expensive option, especially by comparison with the optimistically projected cost of nuclear power and heavy fuel oil, which was then very cheap.

As in many other countries, a significant expansion in the role of nuclear power was therefore planned in Britain from the 1960s onwards. This was based on forecasts of high electricity demand growth and low nuclear costs, both of which now appear difficult to justify. But errors in technology choice and procurement strategy meant that for much of the 1970s, nuclear power was not a viable option in Britain. In addition, the ample British reserves of coal, oil and gas (compared to those of France Germany and Japan) that were by then apparent, meant that the risk of interruptions to international fuel supplies was not such a strong policy-driver in Britain as in countries with few indigenous resources. Together, these circumstances meant that the plans for nuclear expansion were not fulfilled. By the time of privatization, nuclear power still represented only about 20 per cent of power generation in Britain with most of the plants having been ordered in the 1950s and 1960s.

While the continued low contribution of nuclear power to electricity supplies largely resulted from policy failures, the limited contribution of oil was more a policy choice. Both the 1956 Suez Crisis and the 1967 Arab–Israeli War could have ended in conflicts which engulfed the main oil-producing region. But these apparent threats to the security of supply of oil from the Middle East were seen as only temporary problems while hostilities continued, not requiring significant changes to policy. Nor was the depletion of world oil recognized as a significant argument against the use of oil for power generation.

The main arguments against a large-scale shift to the use of imported oil in power stations were the detrimental effect that the substitution of an imported fuel, such as oil, for indigenous coal, would have had on the balance of payments and the impact on mining communities of reduced coal usage. Such a substitution would also have been inconsistent with Government coal policy which, since 1950, had seen massive investment in the deep-mined industry to expand and mechanize production, concentrating output as far as practical on pits with long-term economic potential. The amount of oil-fired generating capacity was therefore limited by the Government in the 1960s and was usually confined to sites adjacent to oil refineries where the low-value residual fuel oil could be supplied cheaply.

By the early 1970s, compared with the situation in the mid-1960s, British coal had become more competitive as production was concentrated more on the lower-cost pits. However, industrial disputes involving the main miners' union, the National Union of Mineworkers (NUM), began to occur after a long period free of major strikes. In the two years before the 1974 Oil Crisis, the Government sanctioned the ordering of nearly 6000 MW of oil-fired plant. Some of this capacity was ordered ahead of need to provide work for the British heavy electrical industry but the Government and the CEGB then expected oil to be the cheapest, long-term non-nuclear generation option available despite the delicate state of international relations in the Middle East. How far industrial disputes in the British coal industry played a role in the decision to order and complete this plant is not clear.

The 1974 Oil Crisis meant that oil was no longer a realistic option to take a significant share of the generation market. The immediate need to substitute for oil in power generation that followed the Oil Crisis led to increased UK power-station demand for coal. Limited mining capacity constrained production and meant that this potential demand could not be immediately fulfilled. Nevertheless, the expectation of continuing rises in the already high world fuel prices meant that the future of British coal seemed assured. As a result, coal-miners and managers increased their real earnings by appropriating most of the large competitive advantage given to British coal by the dramatic increases in oil price. A National Coal Board document, *Plan for Coal*,[1] which was endorsed by the Government, forecast a significant increase in the market for British mined coal based on direct use of coal in industry and its use to meet expanding electricity demand. Neither source of additional demand materialized.

A second Oil Crisis, resulting from the Iranian revolution in 1979, was short-lived and gave extra impetus to a new competitor to British coal, the rapidly growing international sea-borne trade in coal. This trade was based on low-cost coal mined in countries such as the USA, Australia, Colombia and South Africa. These countries had little of the geopolitical risk associated with oil and had much more accessible and richer coal reserves than the UK. However, Government did not allow any significant increase in coal imports until electricity privatization.

Natural gas was discovered in the North Sea in 1965 but it was not a feasible option for power generation until privatization. Despite early attempts by the ESI to obtain supplies for power station use, the UK Government, supported by the nationalized gas industry, decided that the reserves were limited and better used directly to substitute for gas manufactured from coal to provide a cheap and clean fuel source for industry and households. In Britain, as elsewhere, natural gas was seen as a premium fuel, too valuable for bulk use in power stations which could use much lower-quality, cheaper fuels. This impression was reinforced by the first Oil Crisis, when natural gas became closely associated in the public mind with oil, and a perception of limited reserves and potentially volatile world market prices was created.

Pressures for Change

In most countries, after the two oil shocks, dependence on indigenous supplies of fuel would have been seen as a positive contribution to security of supply. In Britain, a polarized political climate and the coal strikes of 1971/72, 1973/74 and 1984/85 meant that indigenous coal has been regarded by Conservative Governments (particularly those led by Mrs Thatcher) more as a security liability than an asset. This record of poor industrial relations and the psychological effect of the Oil Crises ensured that avoidance of over-reliance on one particular fuel source (especially British coal) became an important stated element of national energy policy in the 1980s. However, it is difficult to find concrete examples of this objective underlying significant investment decisions. The apparent pursuit of diversity has been used as a cover for a more politically sensitive objective. Government Ministers of the time have subsequently admitted that the diversity justification for the 15 GW Pressurized Water Reactor (PWR) programme announced by Mrs Thatcher in 1979 and vigorously defended by the Government until her downfall was less important than its contribution in weakening the power of the NUM.

Despite pressures within the ESI at various times from the 1960s onwards to increase use of oil, nuclear power, imported coal and natural gas, British coal maintained a dominant share of the UK power generation market right through to the time of privatization.

The main area of concern in fuel purchasing at the time of electricity privatization was therefore the British coal industry. By then, world fuel prices were low and stable, and the perceived risk of interruption to international fuel trade was small. The defeat of the 1984/85 miners' strike meant that industrial relations problems in the British coal industry were no longer a realistic threat. The increase in confidence in world fuel, especially coal, markets meant that the perception was growing that the strategic advantages of maintaining a British coal industry, even one free of industrial relations problems, were not worth the apparent price premium that was being paid for indigenous coal.

Experience Post-Privatization

Subsequent to privatization, a growth in demand for natural gas for power generation has forced the reduction in demand for British coal that the Government hoped for. However, CCGT technology was largely unknown in the late 1980s in Britain and conditions in the market for natural gas did not come into the thinking on policy for fuels for power generation prior to privatization. There was little pressure to maintain the ban on use of gas for power generation that had been imposed by the European Commission after the Oil Crisis, but there was equally little expectation that gas would become a major power-generation fuel. There was some expectation that oil, or oil derivatives such as Orimulsion,[2] would increase in significance. However, it was coal purchased on the international market that was expected to profit most from privatization. At the time of privatization, there seemed to be a range of fuels available on world markets to the ESI with secure long-term availability and low prices. Concerns about fuel supply security were therefore not a major consideration.

In the post-privatization regime, for generators to be successful in the day-to-day price competition that it was envisaged the Pool would provide, they must minimize short-term costs and retain flexibility. This gives a strong advantage to technologies like the CCGT which has low fixed costs and variable costs which can be readily reduced if the plant is not used. As a result, the cost structure of any capital-intensive technology, be it nuclear power or large-scale renewables such as tidal power, is too inflexible to be attractive in the new structure. Fuels which can only be developed if they can be contracted long-term at fixed quantities and prices, such as British coal, are also unattractive in a system based on short-term competition. It should therefore be no surprise that, as argued in Chapter 9, British coal and nuclear power have been the main casualties of privatization to date and the prospects for large-scale renewable projects are similarly poor in the absence of subsidy.

Assessment

The period since electricity privatization has, as the Government hoped, been one in which international fuel markets have remained stable and prices low. The volume of international trade in coal has grown significantly but with much new mining capacity coming on-stream in a number of countries, this has not adversely affected the price and availability of coal. The international supply of oil has remained secure despite the Gulf War in 1991 and there is little sign that the existing over-capacity in oil production will be quickly absorbed. The international market for pipeline gas has also developed strongly in Europe partly based on new demands for natural gas as a power-station fuel, using increased supplies from Algeria, Norway, the Netherlands and the former Soviet Union. Additional pipeline and liquefied natural gas (LNG) supplies from countries such as Nigeria and the Gulf States are planned, widening even further the resource base Europe can draw on. A large gas pipeline from Britain to Belgium, expected to be completed in 1998, will give the British producers and consumers the opportunity to buy from, or sell gas to, this growing market.

Overall, the wide range of cheap fuels available and the stability of markets have meant that conditions in international fuel markets could not have been more favourable to the new privatized system. It may be that the market has learnt from the various sharp movements in international fuel markets, the price rises of 1974 and 1980 and the price collapse of 1986, and that world markets are now more robust to such shocks. However, the Middle East remains politically unstable, particularly in its attitudes to the West, and major and potential gas suppliers such as Russia, Algeria, Iran and Nigeria face serious internal divisions which might adversely affect their export capability.

If gas became too expensive or insecure, the obvious alternative would be coal. It now seems likely that, as argued in Chapter 6, only a limited and declining deep-mined coal capacity will exist in Britain after the expiry of the current coal contracts in 1998. Increasing coal production in Britain would therefore need major investment programmes to develop new capacity through new seams and new pits. Without long-term commitments of a decade or more to take the output of this new capacity, financing it is unlikely to be possible. Particularly if the generation market becomes more competitive and competition in supply to domestic consumers is introduced, no company will have a secure enough market-base to make such a commitment. Greater use of imported coal might raise politically awkward questions about why indigenous coal was not being used, but these can probably be overcome and with some investment in import facilities, imported coal would not be too difficult an option. Whatever the source of coal, it is likely that the technology used would have to offer significantly better environmental performance in terms of acid and greenhouse gases than current technology.

Equipment Supply[3]

The Old System

Most of the equipment used by the ESI is supplied by the heavy electrical industry. This is a very unusual industry with much of the equipment supplied by it being characterized by demanding manufacturing processes, a need for heavy expenditure on research and development, very high unit-product values, long lead-times and product lives, and highly cyclical demands. As a result of these conditions, only very large, usually diversified companies have had the technical and economic strength to remain in the market.

Ability to supply such equipment was regarded by Government as a valuable strategic asset in the few countries which could sustain such an industry. A heavy electrical industry was also seen as a demonstration of national industrial capability and as a method of nurturing advanced manufacturing and design skills. From a macro-economic point of view, purchasing such expensive equipment from an indigenous supplier had an important positive effect on the national trade balance and successful exports of such equipment were expected to pave the way for exports of other goods.

In accordance with their statutory obligations to maintain a secure supply of power, the ESI in the UK, as elsewhere, has always placed an extremely high value on the reliability of its purchases. A reputation for quality has therefore often been of comparable importance to price in choice of supplier. Purchasing from a national supplier which the utility dealt with frequently meant that the utility could expect preferential treatment in the supply of new equipment over less-valued customers. This reduced the risk that supply security would be jeopardized by failure to acquire equipment in time to meet expanding demand. This priority also extended to servicing existing equipment and this was expected to reduce down-time for key items of equipment, again improving the security of electricity supply. Utilities could also exert pressure on home suppliers to customize designs to their own particular requirements. The value attached to priority in dealing with orders, customized designs, and to reliable maintenance and repair meant that bid prices were seldom the central factor in choice of equipment supplier.

As a result of these conditions, the Governments and utilities of countries with such industries were wary about any actions which damaged the prospects of the equipment supply companies. De facto bans on avoidable equipment imports were therefore the rule in most of the handful of countries which still had an equipment-supply capability after World War II. Protected home markets made up about 80 per cent of the world market. The comparatively small volume of equipment traded on the international market meant that most equipment suppliers were heavily dependent on their home markets, which also served as a 'test-bed' and 'shop-window' for their new technology.

These considerations meant that the equipment supply industry could expect to win all the available orders in its home market. For export markets, once a supplier established a strong relationship with a utility, that supplier could expect to win most of the orders the utility placed at a premium price. In these circumstances, there was little incentive for suppliers to compete strongly against each other in export markets. How far these rigid equipment supply patterns were reinforced by the existence of the long-established international cartel to regulate export prices, the International Electrical Association, is difficult to determine.[4] The result of these conditions was that there were high technological and commercial barriers to entry for new equipment suppliers and that equipment procurement patterns were rigid. The development of nuclear power technology, with its more stringent technological demands and its military sensitivity, reinforced these effects.

In Britain, prior to privatization, the equipment supply industry developed new products in partnership with the ESI. In return, all the orders placed by the British ESI were allocated between the British companies at prices set by negotiation rather than competitive bidding. In the 1960s, the Government began to intervene more explicitly and regularly in the heavy electrical industry and its relationship with the ESI, as part of an active industrial policy. The Government tried on several occasions to encourage mergers amongst the handful of British power-plant suppliers. Its motive was to create world-class companies, especially in high-technology areas. This would bring economic advantages to British electricity consumers and would increase the competitiveness of British equipment suppliers on international markets. Government also required the ESI to place orders ahead of need on a number of occasions purely to maintain a steady flow of orders with the equipment suppliers.

Pressures for Change

There was concern that the prices paid by the ESI in Britain were unjustifiably high and, particularly for nuclear power technology, that the equipment purchased was inferior. There was also criticism that the CEGB interfered too much in the design process, duplicating many of the equipment supply companies' tasks. Proponents of privatization claimed that market forces would ensure that costs were minimized and inefficient procurement practices would be forced out.

There were also forces external to the UK which were exerting increasing pressure to change the old practices regardless of how the ESI was organized. The European Commission and the General Agreement on Trade and Tariffs (GATT, now the World Trade Organization) were becoming more powerful and were both pursuing policies which would encourage free trade in goods. The policy of favouring national suppliers instead of placing orders on the basis of competitive bidding ran against their basic free-trade philosophy. Previous trade liberalization initiatives by these two organizations had been opposed by the equipment and electricity supply industries, and utility procurement had been exempted from the measures implemented. However, the Commission's Procurement Directive, finally ratified in 1993, and the Uruguay Round of GATT negotiations, completed in 1994, both targeted utility procurement for reform and have required orders to be placed on the basis of free and open international tender.

However, apparent inefficiencies in the relationship between the equipment supply industry and the ESI were a much less significant force driving electricity privatization than were those in fuel supply. The CEGB itself was showing impatience with the convention that it should spread its orders evenly amongst the British suppliers and was trying to encourage the British companies to compete more vigorously against each other. However, the option of purchasing standard equipment on the international market was not seriously pursued.

By the time electricity privatization was undertaken, the structure of the heavy electrical industry was also changing dramatically, away from one based on 'national champions'. Two of the major companies, Brown Boveri (Switzerland) and ASEA (Sweden), merged in 1987 to form ABB, quickly followed in 1988 by a merger of the heavy electrical sectors of the French company Alsthom and GEC (UK) to form GEC-Alsthom. These two new alliances took over a number of smaller companies in Italy, Germany, Eastern Europe and the USA, dramatically changing the face of the equipment supply industry.

From a situation in which six or seven countries could all choose from two or more apparently independent, nationally based suppliers, the industry was transformed, during the period privatization was carried out, to one dominated by only five, mostly internationally based suppliers or groups of suppliers. Three of these are based in Europe, Siemens of Germany as well as ABB and GEC-Alsthom. The other two competitors were US–Japanese alliances, Westinghouse with Mitsubishi, and GE with Toshiba and Hitachi. The market for equipment supply, by becoming more concentrated and international, had become difficult for Governments to influence even before privatization was completed.

In some respects, the changes were somewhat less dramatic than they might appear at first sight. The two US–Japanese alliances and Siemens were little changed by the process. In addition, even before the restructuring, the industry was already highly concentrated given that many of the national suppliers were dependent on technology licences from the major players. Nevertheless, as France demonstrated in the 1970s with its nuclear power programme, even weak national companies can be built up to become strong competitive companies given strong political backing and a sufficient volume of orders. In countries such as Italy and France, the policy of using state and utility funds to develop a national technology resource became more difficult to sustain because most of the national companies had been absorbed into bigger groupings with no particular national allegiance.

This market concentration is well illustrated by gas-turbine technology, which is currently the most important capability in the heavy electrical industry. There are only four independent designs of large gas turbines in the world, three of which are those of Siemens, ABB and Westinghouse/Mitsubishi. The fourth is that of GE which is licensed to Hitachi, Toshiba and GEC-Alsthom. The large numbers of other suppliers of large gas turbines are all licensees or manufacturing associates of these four groupings.

Experience Post-Privatization and Assessment

In the UK, these changes to the heavy electrical industry removed a potential cause of conflict for the restructured ESI. It would have been unwilling to shoulder the responsibility of supporting weak local suppliers and funding new product development. The whole ethic of the new industry was based around market forces, and any process of supplier choice which did not involve open bidding would have been unthinkable. Electricity companies could not maintain a special allegiance to British suppliers if this did not minimize short-term costs because of the advantage this would hand to their competitors in the electricity market. The ESI was also confident in the strength of its management to extract the best possible terms from suppliers. Early experience with substantially lower prices for equipment and conspicuously strong technological and price competition in the supply of gas turbines seems to have borne out this confidence.

However, this strong competition may not have extended across the range of products and may have been no more than the newly established equipment supply alliances indulging in short-term jockeying to establish a strong position in the new structure. The record of the sector in behaving competitively is poor and the world heavy electrical industry is now too concentrated for it to be assumed that competition will be intense. It may be that when the new internationally based structure has settled down, perhaps after further mergers, and opportunities for technological development are more risky and expensive, the need for Government or utilities to intervene for strategic reasons in the market to stimulate competition and develop new technologies will re-emerge.

Technology Choice and Development

The Old System and Pressures for Change

Generally, Government involvement in ESI technology choice and development was based on the premise that the ESI, through its R&D activities, would carry the national interest in technology development. This was seen to require that new technologies should be developed within national boundaries so that they were tailored to the country's needs and were freely available to the local ESI. Further considerations were the creation of future export capabilities and avoidance of dependence on foreign suppliers for strategically important technologies, such as nuclear power plants and the associated services. The issue of concern examined here is whether it will be important in the future for the British Government to influence technology development for the electricity supply industry and how it might do so with a privatized ESI.

As a result of the implicit duty of the nationalized ESI to carry the national interest in technology development, it was encouraged to pursue a technology development policy with a much longer time horizon than would be typical for a private-sector company. This was particularly noticeable with nuclear technologies. In Britain, the CEGB funded conventional nuclear technology, fast reactors, renewable technologies and advanced coal-combustion methods.

To a large extent, the CEGB acted as an agent of Government in commissioning and carrying out long-term R&D on new electricity technologies. The UK Government also carried out and sponsored complementary research programmes. This resulted in the conduct of a large volume of R&D, but there is room for debate about the priorities chosen and the effectiveness of some of the long-term programmes chosen. This is despite the fact that the CEGB and other nationalized-industry R&D programmes were all vetted and approved by the independent Advisory Council on Research and Development which advised the Minister in relation to his statutory duty to ensure the volume and relevance of public-sector R&D.

There was concern over the management of the programme, particularly whether public funds (taxpayers' money and electricity consumers' money) were being spent where private-sector equipment supply industry money should have been spent. There were also worries that the public sector was carrying a risk of failure that should also have fallen on the private sector. The choice of technologies supported was seen to be based partly on internal vested interests within the CEGB and partly on how interesting and challenging the R&D would be.

Nevertheless, the publicly owned ESI had large R&D programmes approved by successive Ministers as being in the public interest, and major R&D and engineering laboratories with a high international reputation.

The CEGB felt it necessary to carry out this R&D because its private-sector suppliers showed little inclination to do it, even though the CEGB was required under the Government's 'Buy British' policy to place any resulting orders with them.

Internationally, the record of success in public spending in energy technology development, directly by international agencies and Governments, and through electric utilities, is generally rather poor. Large amounts of money have been spent on developing nuclear technologies, such as fast reactors, that have not achieved commerciality and on fusion power, a technology which even its proponents do not expect to be viable within the next 50 years. Other technologies which have received significant Government support, including clean-coal technologies and large-scale renewables such as tidal power, have long been under development and also show little sign of any commercial breakthrough.

Indeed, there has been a consistent tendency for public-sector R&D to focus on what are now regarded as less promising large-scale technologies such as large wind turbines, tidal power and geothermal energy. More commercially attractive small-scale wind and biomass plant has, for the most part, only been developed in the private sector. Even the gas turbine, which is the key item in CCGT technology, and which is often portrayed as being developed independently by the heavy electrical industry has benefited from the vast Government sums spent on developing jet engines. Given this poor record, the pressure to abandon or dramatically scale down any Government involvement in technology development is easy to explain.

Experience Post-Privatization and Assessment

The magnitude of fossil fuel resources is highly uncertain and the timing of the need for replacement energy sources correspondingly unsure, but the resources are finite. The use of fossil fuels involves significant environmental impacts which it might not be possible to ameliorate, global fossil fuel supplies remain vulnerable to disruption and the private sector is reluctant to invest large sums of money in technologies that may only offer a return after decades. These factors suggest that rather than simply abandoning long-term technology development strategy, a more far-sighted response may be to develop better strategies.

The loss to Governments of the nationalized industries as levers of policy has brought the problem of how to develop appropriate new electricity technologies to the fore, but this issue has not arisen simply because of privatization. The globalization of the equipment-supply companies means that developing technologies through national companies is now seldom an option; the scale of resources needed to develop new technologies is often beyond the means of single nations and Governmental efforts to develop new technologies must increasingly be carried out on an international basis, for example through the European Union; better ways of choosing the technologies to be supported by public money have to be found; and a more equitable balance has to be struck in allocating the risks in developing new technologies, so that the private sector bears a larger, more appropriate share.

If the arrival on the market of the gas turbine as a mature technology is a demonstration that the equipment supply industry alone can be relied upon to provide the future stream of technical innovations which will drive the electricity supply industry forward, these issues may not be important. However, if as seems more likely, the gas turbine was a 'one-off' born of a particular combination of circumstances, resolving these problems will be a priority.

Nuclear Power Policy

The Old System

The most conspicuous area of Government involvement in the technology policy of the electric power sector has concerned nuclear power. Government involvement in nuclear power policy has already been touched on above and in considering fuel and equipment supply, but the extent of the involvement is so great that it deserves to be considered separately.

As a result of strenuous postwar efforts to acquire military nuclear capability, Britain was one of the first countries to become involved in developing civil nuclear power. Many of the facilities used for military applications, such as Magnox reactors and spent fuel reprocessing plants which were used to produce weapons-grade plutonium, could be adapted for use as part of the power production infrastructure; and the supply of plutonium reprocessed from Magnox reactors was essential for the weapons programme.

The CEGB was initially sceptical about the economics of nuclear power and ordered its first plants at Government instigation. These civil nuclear power plants used the Magnox technology developed by the military sector. The history of the subsequent nuclear power decisions was covered in Chapter 7, but in 1965, 1973 and 1977, decisions on technology choice were made by the Government, not always with the approval of the CEGB. Even electricity privatization and restructuring which, it was claimed, would free the industry from Government pressures, was designed to ensure a major continuing role for nuclear power.

Pressures for Change

Nuclear power R&D was carried out at immense public cost in Britain and elsewhere, almost to the exclusion of other, often more promising technologies. Nuclear power, more than any other civil technology, can only be smoothly implemented if Government provides strong support through enabling decisions on the siting of facilities, access to technologies with strong military sensitivity and guarantees on a range of costs including accident and waste liabilities. But the high hopes of the 1950s that nuclear power would provide a cheap, effectively limitless source of power were not fulfilled and the returns on the large amounts of public money pumped into the technology seem low.

For many reasons, nuclear power was unlikely to profit from the restructuring in Britain of the ESI. Its poor economics, the high capital-intensity, the long lead-times and the high level of user-skills required all strongly mitigated against its adoption. In addition, the requirement for Government involvement in taking enabling decisions and providing guarantees was against the spirit of the reforms which demanded that Government withdraw from ESI decision-making. From a practical viewpoint such support was electorally unpopular and the guarantees potentially very costly.

Experience Post-Privatization, and Assessment

The nuclear power sector had to be withdrawn from privatization at the last minute. To deal with the poor economics of the nuclear power plants and the unattractiveness of nuclear power to private investors, new forms of Government intervention were introduced. The non-fossil fuel obligation (NFFO) was intended to force the industry to invest in new nuclear power plants and the fossil fuel levy (FFL) was an open-ended consumer subsidy introduced to compensate the nuclear generating company for its losses. The NFFO could not achieve its original aim and now merely ensures that existing nuclear plant is fully utilized and the European Commission required that the FFL be phased out by 1998.

Neither of the large privatized generating companies has shown any interest in nuclear technology and Government is withdrawing from its technology development commitments as quickly as it reasonably can. The Government now plans to try again to privatize some of the nuclear power plants in 1996. However, plans to build further nuclear power plants were finally abandoned in 1995 and privatization, if successfully carried out, will not lead to a revival in nuclear power's fortunes.

Nuclear power, in anything like its present form, simply appears incompatible with a privatized, fully competitive system no matter how adverse the developments in fossil fuel markets are. There can also be little confidence that public acceptance of the activities that comprise the nuclear power infrastructure will increase sufficiently to remove the severe problems in siting facilities. It is hard to envisage new nuclear technologies which harness the huge potential scale of the nuclear resource but which do not bring with them these problems. The history of nuclear power is such that Government, utilities and equipment suppliers will be very wary about investing further resources in such a risky venture.

Few countries are still pursuing a policy of actively encouraging an expansion of the role of nuclear power and, given the extraordinary demands designing, building and operating nuclear power plants imposes, it would not appear appropriate for Britain to be investing significant sums in new nuclear technology.

MACRO-ECONOMIC AND SOCIAL POLICY

The Old System

Decisions on whether to import fuel or equipment or to supply them from indigenous resources can have a significant effect on national resource allocation. As discussed previously, the relevance of such decisions to the national macro-economic situation has prompted Government interest in them in the past. The other area of Government influence, prompted by macro-economic considerations, has been the setting of electricity prices to achieve economic objectives, for example on inflation. A further possible cause for intervention, not pursued in Britain, is to hold tariffs down for low-volume domestic consumers as a measure to alleviate poverty. The issue examined in this section is the circumstances which might lead a future British Government to intervene in the ESI in the interests of macro-economic or social policy and what, if any changes would be necessary to enable it to do so.

In the UK, from the time of nationalization, the overall level of tariffs was a direct concern to Government since these largely determined the amount of money the ESI paid to or drew from the Treasury, and the rate of return it achieved on nationalized industry assets. Over this period, the Government's policy has been that nationalized utilities should earn broadly the same rate of return as comparable private sector companies. In addition to this normal shareholder interest, Government intervened in tariff-setting on various occasions for two other reasons: first, tariff increases were delayed or reduced to combat general price inflation: and second, prices for very large consumers were held down and sometimes subsidized to improve the international competitiveness of such companies. These measures were operated under a regime that aimed to preserve the managerial autonomy of the public corporations within broad financial and economic guidelines laid down by Government and within certain statutory requirements to secure the public interest.

In other countries, such as Italy, tariffs for low-volume domestic consumers have been held down to provide cheap power to poor consumers. This approach has not been followed in Britain, and, particularly in recent years, 'fuel poverty' has been addressed by targeted measures such as extra welfare payments in periods of exceptionally cold weather and investment in energy-efficiency improvement measures, such as loft insulation.

Pressures for Change and Experience Post-Privatization

Government involvement in tariff setting is often portrayed as a prime example of the negative impact of Government interference in ESI decision-making. Selective help to groups of consumers such as low-income households and electric-intensive industry, or smoothing large price adjustments, for example after a major increase in fuel cost, is often opposed by the other groups of consumers or the tax-payers, who must foot the bill. Those opposed to Government intervention argue that these measures are ineffective in combating the underlying problem, be it poverty or industrial competitiveness. For example, they argue that suppressing price rises provides only short-term benefits which are more than lost in the longer term when price rises are allowed to catch up the lost ground.

The restructuring of the ESI has changed, rather than reduced, the scope for Governments to influence electricity tariffs. Through the Government-appointed regulator, the Director General of Electricity Supply, it can have a powerful influence on the level of tariffs for the monopoly parts of the service. Now that Government is no longer the owner of the industry and recipient of its profits, forcing these prices down no longer leads to a risk of reduced Treasury income.

It is now difficult to justify tailoring tariffs to meet other policy objectives, such as increasing the competitiveness of electricity-intensive industry or alleviating poverty, especially to the European Commission. The alleviation of poverty is usually seen as more efficiently addressed directly through taxation and social-security measures, and by targeted programmes, such as loft insulation or replacing inefficient, expensive heating systems, rather than indirectly by cross-subsidizing power.

It is also difficult to argue for electricity cross-subsidies to electricity-intensive industry in today's international trade regime. Subsidies are usually only suggested as a retaliatory measure against countries which are suspected of subsidizing their own industries. However, there is still ample evidence that, internationally, much electric-intensive industry still is treated favourably. Even in Norway, where the electricity supply industry was liberalized along similar organizational lines to Britain in 1990, electric-intensive industry was allowed to sign long-term, low-cost electricity supply deals which insulated it fully from the impact of the new electricity market.

Assessment

The six years since privatization has been a period in which there would have been little pressure, under the old nationalized system, for the Government to intervene on macro-economic grounds in electricity prices. Inflation has been low by comparison with the previous three decades, fuel prices have been stable and low, and there has been considerable momentum to free-trade policies. Unemployment has remained persistently high, but reducing it has not been such a high priority as it would have been for the administrations before those of Mrs Thatcher.

At present, low inflation, low fuel prices and the significant scope for efficiency improvements (the 'X' in the RPI–X formula) mean that electricity prices are falling in real terms and are stable in nominal terms (not adjusted for inflation). However, if fossil fuel prices were to rise and cause higher inflation, the situation could be very different. Fuel price increases would be passed on in full to consumers. The 'RPI–X' formula, under which the monopoly services are priced, would ensure that electricity prices largely kept pace with inflation especially if the scope for efficiency improvements (a large 'X' factor) was small. In this situation, electricity prices could be seen as an 'engine' of inflation which the Government might feel obliged to control more directly by changing the regulatory formula or by requiring the regulator to apply additional 'one-off' measures.

Similarly, the dominance of free-trade policies cannot be assumed to continue. If full employment becomes a central Government objective in a significant proportion of the developed countries, the free-trade regime might be difficult to sustain and countries may be drawn into trade wars. In this situation, there would be considerable pressure, given the existence of (or suspicions of) electricity cross-subsidies, to retaliate with new subsidies to industrial consumers.

It is more difficult to envisage strong pressure to subsidize domestic electricity prices. While there can be doubts about whether sufficient resources are now being devoted to alleviating fuel poverty in Britain, the general approach of targeting resources at specific groups of consumers and encouraging energy efficiency seems a better way of addressing the problem than do subsidies.

ENVIRONMENTAL PROTECTION

If there is disagreement over the extent to which there is a legitimate role for Government regarding other areas of strategic decision-making in the ESI, there is wide agreement that Government must set and enforce environmental protection standards. The weight of this responsibility has increased greatly in recent years with the parallel concerns over acid rain and global warming. Government – increasingly via the European Union rather than at a national level – has a duty to set a policy framework to ensure that environmental pollution from all sources is reduced to levels which the environment can tolerate on a long-term basis and to require that polluters pay the cost of any clean-up. The issue examined here is how privatization has affected this duty.

The electricity supply industry has always been, and is likely to continue to be, an obvious instrument for the Government to achieve environmental objectives. Imposing measures on the ESI does not affect personal behaviour in the way that, for example, restrictions on car use would, and the economic impact of such measures can often be effectively concealed within the overall cost of electricity supply.

Over recent years, the emphasis on the implementation of Government environment policy has shifted with privatization to greater use of market-based instruments as opposed to the traditional 'command and control' measures which prescribed, to a much greater extent, the technological means that would be used. This trend is apparent worldwide in parallel with, but independent from, liberalization and privatization. Although often not acknowledged, both approaches involve Government intervention: it is simply the manner, rather than extent, of intervention which distinguishes them.

'Command and control' mechanisms and market instruments both have their failings but there is little doubt that Government will continue to have a responsibility to set standards and devise policy instruments which achieve the required environmental standards at the minimum financial cost to consumers. The specific measures now used for environmental regulation are discussed in detail in Chapter 5.

The period since privatization has been one in which the procedures have been little tested. The large volume of new CCGT plant and the increased output from the nuclear power stations, both little connected with any environmental concern, have meant that the Government's targets on acid and greenhouse gases have been easily met with no need for Government intervention to force companies to carry out additional investments.

If, in the future, even tougher targets on, for example, acid and greenhouse gases, are agreed internationally, it is less likely that it will be possible to meet them without imposing new requirements on the electricity supply industry. Provided the costs can be passed on to consumers, and that the competitive position of, for example, a particular generation company, is not adversely affected, enforcing such decisions should not be too difficult. The process will involve negotiations between the companies, the regulators (environmental and economic) and Government, and Government will have to be vigilant to ensure that private companies do not pass on excessive costs to consumers.

PROMOTING ENERGY EFFICIENCY

The promotion of energy efficiency has been seen by many as a high priority since the first Oil Crisis. The benefits of using energy efficiently include reductions in the environmental impacts of electricity supply, reduced vulnerability to international fuel markets, alleviation of so-called fuel poverty and more economically efficient use of national resources. Despite wide recognition of these benefits, the British Government, like most Governments, was frequently criticized for not intervening more strongly, before privatization, to ensure that the economically attractive options were taken up. This section examines briefly why this occurred and how privatization has changed the outlook for energy efficiency.

The basic problem appears to be the imbalance between the strength of the supply and demand sides of the electricity system. Considerable Government effort was spent ensuring the survival of indigenous fuel and equipment supply industries, but little was done to ensure that electricity was used efficiently. This imbalance can be explained by the fact that equipment and fuel industries are large and powerful, and stand to gain considerably from maintaining electricity demand as high as possible. Consumers of electricity have often not been aware of the scope to reduce electricity demands profitably. Electricity bills are not usually a major item of expenditure and consumers will tend to devote their resources to meeting their major items of expenditure at lowest cost. Poor households do tend to spend a larger proportion of their income on electricity, but do not have the money to invest in energy efficiency or they live in rented accommodation that they cannot alter – they suffer from 'fuel poverty'. The industries involved in energy-efficiency services are relatively weak and fragmented, and not politically influential. As a result of these factors, deriving effective policies for energy efficiency, worldwide, has proved difficult and there is no consensus about the best approaches.

However, the British Government did not seem to place a high priority on this area prior to privatization. From 1967 to 1989, the electricity supply industry was allowed to make investments that made only a 5 per cent real return on capital, a figure that was then raised to 8 per cent (see Chapter 2). By contrast, there was ample evidence that, for a variety of reasons, users of electricity were not taking up energy-efficiency investment opportunities which offered a far better rate of return. Such a misallocation of resources would have tended to reduce national wealth, for example by making manufacturing industry less competitive in international markets and by reducing the amount of money domestic consumers had to spend on other goods.

Privatization can be seen to have addressed half of the problem, the overemphasis on the supply side. The Government pre-occupation with indigenous fuel and equipment industries has largely disappeared and, by privatizing and introducing competition to generation, much of the electricity supply industry investment has to compete with other national demands for capital and must achieve as good a rate of return (not less than a real annual rate of 12 per cent), adjusted for perceived risks. However, measures on the demand side are still inadequate and even relatively non-controversial measures are not being applied. Appliance labelling, by which consumers could compare the efficiency of different appliances, has only slowly been introduced. Market instruments could also be used and, for example, standards could be set for domestic appliances, which the appliance industry would have to find the best way of meeting, in the same way as the Government sets environmental standards. But the British Government has favoured voluntary approaches rather than the mandatory ones adopted in some US states.

If these types of measure are not sufficient fully to correct the imbalance, it is difficult to imagine further options which do not involve more active involvement, of the type the Conservative Government has tried to avoid, in the electricity supply industry. For example, in the USA, electricity prices have in some cases been increased more than was necessary, to generate funds for demand-side management programmes which would improve efficiency. Provided consumers find that the value of the electricity saved is larger than the tariff increases, they should, at worst, be indifferent to these price rises. There are important practical problems with such programmes, such as estimating the savings attached to efficiency measures and ensuring equity, so that a high proportion of those facing the tariff increases actually do derive some benefit.

While energy prices are low, the economic penalties of not taking up energy efficiency opportunities are reduced, but low prices cannot be assumed to continue, nor can supply security be guaranteed. The environmental benefits are not likely to diminish and, as reducing greenhouse gas emissions becomes a higher priority, the importance of energy efficiency will be further emphasized.

Corporate Strategies in the ESI

Up to the time of privatization, the electricity supply industry appeared to be a largely homogeneous, engineering-based industry involving the construction, operation and maintenance of long-lived, complex, technologically demanding equipment. The decision to separate the industry into four distinct components and to change the way in which it conducted its business has begun to destroy this perception of homogeneity.

The concept of, and rationale for, de-integration were unfamiliar at the time of privatization. It then appeared that the Government was aiming at a structure in which the two monopoly businesses, high-voltage transmission and low-voltage, local distribution, would be carried out by companies with a specific mission to do this task. The competitive elements, generation and supply to final consumers, were to be opened up so that a large number of companies could compete in these sectors. Company links between the four sectors, vertical integration, were portrayed as undesirable because of the risk of cross-subsidization, for example, of the competitive businesses by the monopoly businesses.

A logical conclusion to such reasoning might have been a structure in which a large number of companies were involved in generation, a large but different set of companies was involved in supply, and a third set of companies with geographical monopolies would supply distribution and transmission services. Under this structure, it was assumed that competition would mean that no specific regulation of the generation and supply businesses would be needed. In order to regulate the companies operating the monopoly elements of the system effectively, it would be necessary for them to be either single-mission companies or that the accounts for the regulated business be very strictly segregated from those of any businesses in other industrial sectors that the company was involved in.

Such a structure was not immediately achieved, mainly for the practical reasons discussed in Chapter 3. The generation sector was concentrated in only two private-sector companies because of the attempt to protect nuclear power. Distribution and supply were not separated but were the responsibility of the same set of companies, the RECs. In part, this may have been because the Area Boards which were to become the RECs were commercially too weak to be split up at that stage. It may also have been because much of the supply market was to remain a monopoly for the first eight years and that allowing this integration was not seen as an immediate threat to competition.

Ownership of the company that operated the grid, the National Grid Company (NGC), was given to the RECs, because of perceived difficulties in valuing and selling a company with operation of a transmission network as its sole business. Any risk of anti-competitive behaviour by the RECs that ownership of the transmission network raised, was countered by provisions which limited the extent to which they could influence NGC policy. The two large pumped-storage power stations were given to the NGC, probably because of the important role of this plant in ensuring the stability and security of the high-voltage grid.

Some less easily explicable structural decisions also compromised the strict separation of businesses that it appeared was being sought. The RECs were allowed to own as much power-station capacity as they wished, although they were able to contract for no more than 15 per cent of their power requirements from such plants. The generation companies were allowed to compete to supply final consumers in the sectors of the market that were opened up to competition.

Following privatization, the basic structure of the industry was largely frozen for the first five years by the Government's 'golden share' holding in the RECs which gave them right of veto over any proposed change of ownership. However, since 1994, the regulator has begun to act to try to remove some of the 'faults' in the structure. He required the RECs to sell most of their shares in the NGC in December 1995 and he directed the NGC to sell its pumped-storage generation power plants. In order to increase competition in the generation market, particularly in the non-base-load sectors, he required National Power and PowerGen to sell 4000 MW and 2000 MW respectively, of existing coal-fired plant by March 1996.

While the regulator's activities appear to have as their objective the achievement of an 'ideal' de-integrated structure, the direction of Government policy has generally been rather different. It is trying again to privatize most of the nuclear power plants and if this is successful, it will mean that the anomaly of public ownership in an otherwise private system will be much reduced and that the nuclear sector will be more fully exposed to competitive forces. Other decisions by the Government following the burst of takeover activity that has continued since the lapsing of the Government's golden shares in the RECs show little consistency or logic. The decisions that vertical integration in generation and supply was not objectionable in principle, and that horizontal integration into other utility services was acceptable, do put in doubt the Government's commitment to a fully competitive industry. However, the Government sought to justify its decisions not to allow the takeover of RECs by the two large generating companies, and to signal that it would exercise its golden shares in these two companies to block takeovers on the grounds that they could jeopardize the emergence of supply competition in 1998. The Government's failure to send unambiguous signals about the long-term structure and ownership of the industry it wished to see means that electricity supply industry companies will be legitimate targets for any company, whatever its background, wishing to move into this sector. The competition issues raised by these takeovers are discussed in Chapters 4 and 10. Equally, ESI companies are free to move into any business which serves their own corporate interests. This means that the electricity supply industry is likely to be rapidly assimilated into the general industrial structure and that it may be only a short period before an identifiable electricity supply industry based on companies whose main business is electricity will cease to exist.

The four separate components of the electricity supply industry have developed since privatization. This has revealed that, while they are all necessary links in the electricity-supply chain, they are in fact very disparate businesses, which generally do not fit together with any particular internal industrial logic. For example, distribution has been revealed as an engineering-based business aimed at maintaining a complex technological system, while the supply business is increasingly one centring on commodities trading and product marketing.

These disparities will tend to stimulate takeover and merger activity. This restructuring is likely to be driven by two priorities: identification of complementary businesses which can either take over parts of the electricity business or be taken over by electricity businesses; and protection of interests against competitive threat. In order to see how this restructuring might develop, it is useful to examine the nature of the four electricity businesses in more detail.

Transmission and distribution now appear to be engineering-based, requiring skills in contracting, maintenance and system control. The business seems to be low-risk with little scope for growth of the business in the UK except by takeover or diversification. Once the economic regulator gets to grips with these businesses and squeezes out excess profits, profits are likely to be limited but assured. For such companies, the opportunities in, and threats from, complementary businesses appear to come from any infrastructure-delivered service including water, gas and cable services. There may be expansion prospects overseas into similar industries, especially in developing countries where privatization is being introduced and the need for foreign capital allows attractive levels of profit to inward investors. It now seems unlikely that regulation in Britain will be so weak as to allow ownership of the network to be used to advantage in the competitive sectors of the new system. This means that generators have no special incentive to take over distribution businesses and that the distribution and supply parts of the RECs will tend to be de-merged. The demerger of the RECs' two businesses may also be seen by the regulator as a useful step to encourage competition in supply.

The generation-sector business now appears to be one based on operation of continuous-process plant, procurement of fuel and equipment, and management of asset-lives. Privatization has revealed that the capital-intensity of the generation sector makes it very high risk. If competition in generation intensifies, these risks will increasingly fall on shareholders rather than consumers. For the existing large companies, there is little scope for growth of the generation businesses in the UK which would not excite regulatory concern on monopoly grounds, although the aim to make the sector competitive does mean that high profits would not necessarily be unacceptable. Experience of operating in a competitive environment may give the UK companies a head-start in acquiring generation assets in countries where privatization is taking place.

The intrinsically high risk of generation suggests that reducing this risk will be a high corporate priority. One avenue for this would be for generation companies to integrate into supply to final consumers, with the objective of creating a captive market. This would reduce the risk in building new plant and give greater assurance that existing assets would be utilized. This would be easier to achieve for domestic consumers, who are less cost-sensitive and more likely to show supplier loyalty than larger consumers. Experience outside the UK, particularly in the USA and the Netherlands where combined heat and power (CHP) plants associated with continuous-process plant operation, such as oil refining and chemicals, have proved economically very attractive, suggests that this might prove a useful option for generators in building loyalty with industrial consumers.

It is the supply business that has departed most from the traditional engineering image of the ESI. The gradual opening up of competition in supply means that the restructuring the sector is likely to undergo is now at only a very early stage. The uncertainties about whether and when it will be possible to introduce real competition for domestic consumers mean that its final shape is difficult to predict. The business appears to be one of commodities' trading, and marketing these commodities to final consumers. The turnover of the business is very high compared to the profits, but the value of assets required to run such a business is very low and the return on these assets very good.

One clear, general threat to the suppliers is, as argued above, that generators will continue to try to integrate into supply to reduce risk in the generation business. To identify other risks, the supply industry market must be considered in two, largely distinct parts. The high-volume business to large consumers appears to be a high-risk, competitive one with low loyalty, because of the economic sophistication of the customers. The skills required for success in this demand sector are those of a commodities trader. If the duopoly power of National Power and PowerGen over the Pool is broken, there may be scope for futures markets and other trading instruments to be set up which new trading companies will be able to exploit to supply large final consumers. To counter this threat to the existing suppliers, it may be possible for them to increase their customer loyalty by promoting jointly owned CHP schemes with industrial consumers, although, as noted above, this is likely to bring them into competition with generation companies.

By contrast, the nascent supply business to small-volume consumers may be subject to much less risk due to greater customer loyalty. Electricity bills are generally a major item in overall expenditure only for poor households. But this group of consumers, which has the most to gain from exploiting the market to achieve the lowest price, is not likely to be an early target for supplier competition. In part, this is because of the low absolute value of their bills and hence the profitability, but it is also because of practical difficulties, such as high levels of consumer debt, that supply to such consumers sometimes raises. For this low-volume business, other domestic services including gas, water and cable, represent complementary businesses that suppliers might profitably enter.

The practical difficulties of introducing competition for domestic consumers (outlined in Chapter 10), such as the high cost of metering and data processing, are putting in doubt the UK Government's target of introducing competition in 1998. But this should not obscure the powerful position that a supplier of such a service might hold. Access to, and detailed knowledge of, a large, geographically concentrated set of consumers may represent a powerful base from which to expand activities, especially if developments in information technology increase the significance of home shopping.

CONCLUSIONS

The involvement of Government in utility decision-making generally arose for entirely legitimate reasons. However, the original rationale had often been forgotten and much of this involvement was ripe for re-evaluation when the ESI was re-organized. The growth and maturation of international markets in coal, gas and power itself meant that the old response of ensuring security through a high level of self-sufficiency was no longer appropriate. Similarly, changes within the heavy electrical industry, beyond the control of Governments, would have necessitated a new procurement regime. Government technology-development policy, which had long been based almost exclusively on a mixture of nuclear fission and fusion, often within nationally based programmes, had not been conspicuously successful. The scale of resources needed to develop fast-breeder reactors and fusion was beyond the scope of national Governments and the time when they might become economic was judged to be a long way into the future.

However, the period since privatization has been remarkably advantageous to the new structure. Supply security has been maintained and pollution targets met as a result of the large volume of CCGT orders; fuel and equipment markets have favoured the buyers; and the macroeconomic conditions such as low inflation and reliance on free-trade policies have limited the pressure to influence electricity prices. However, it cannot be assumed that these conditions will continue. The fuel and equipment supply industries both have significant structural problems which mean that total reliance on them for the long term may still be risky.

The responsibility for system planning which was previously held by the CEGB has only partly been passed on to NGC and nobody has the task of ensuring there is sufficient generating capacity to meet demand. Already, there are signs that, despite the large volume of plant ordered in the 'dash for gas', the new system cannot be relied on to 'keep the lights on'. The problem is compounded because the peak demands for gas and electricity tend to coincide, and the increasing dependence of the electricity system on gas-fired generation is raising new problems about how to cope with demand peaks.

Electricity privatization revealed two largely unacknowledged facts about the electricity supply industry. First, much of the business involves large financial risks which had previously been borne collectively by electricity consumers and taxpayers. Second, what was previously seen as a largely homogeneous business now clearly comprises a number of disparate parts. It is these two factors, along with the Government's reluctance to set clear guidelines on industry ownership and structure that is causing the current wave of restructuring and will continue to drive this process as the underlying nature of the businesses is revealed. Companies will be seeking to take over others with complementary businesses and those with businesses which can reduce the high level of economic risk that is intrinsic to the industry.

For the monopoly infrastructure businesses, which are not under competitive threat, horizontal integration into other infrastructure industries such as gas, water and other cable services is most likely. For the competitive businesses, vertical integration of generation and supply may be the strongest force, as generation companies seek to reduce the risk of building new capacity and supply companies try to gain an edge over competitors by owning cheap generation sources. However, the supply business is the most unpredictable, because of uncertainties about how competition for all final consumers can be created. The business is relatively small, but possession of it could be a key resource in the future for marketing other commodities.

If the need for strategic intervention by Government does become important, these changes will make such interventions more difficult as structural change in the industry increasingly means that the elements of the electricity supply industry are fragmented amongst diversified, internationally based companies.

Part IV

A MODEL TO FOLLOW?

13

General Conclusions and Lessons

Mike Parker[*]

Introduction

Rather than repeating the conclusions for each of the preceding chapters, this chapter seeks to provide a broad but necessarily tentative assessment of the record of the privatized ESI to date (early 1996). The assessment must remain tentative because numerous elements have yet to be played out. Nevertheless, it is clear that the privatization and associated restructuring of the UK electricity supply industry (ESI) has fundamentally changed the character of the industry. Previously, the industry's management had been dominated by engineers, and the trade unions had considerable influence. Before privatization, the industry regarded electricity as a public service to be universally available, and the monopoly structure enabled full pass-through of costs. Now, electricity is regarded as being no different from other products, and the emphasis is firmly on commercial objectives.

As a result of the changed character of the ESI the 'UK model' of reform of the ESI has assumed much wider significance. The particular feature adopted in England and Wales is the distinction between generation and supply, which are regarded as amenable to competition, and transmission and distribution, which are recognized effectively to be natural monopolies requiring regulation. This distinction between inherently

* Although Mike Parker takes responsibility for the contents of this chapter, he has discussed it fully with the other authors. Consequently, this chapter broadly reflects the views of all the authors.

'competitive' and 'monopoly' activities has thrown a new light, not only on the UK electricity industry, but on the ESI in other countries (and, indeed, on network industries generally). The UK experience therefore deserves careful study. Another main purpose of this chapter is to draw lessons which are relevant for other countries contemplating privatizing and perhaps introducing competition in their own electric power sector.

ASSESSING THE RECORD

Before the privatization and restructuring of the UK ESI, the nationalized industry did not have a good public image, and strong general support for the status quo was not in evidence; but neither was the industry regarded by the public at large as requiring fundamental reform.

The industry appeared to be relatively efficient, making a significant net contribution to public finances, and supply security was high, given that electricity demand was not expected to grow significantly, and that fuel supplies seemed assured following the defeat of the miners' strike in 1985 and the collapse in oil prices in 1986. The impetus to privatization came not from public pressure for improved performance, but from a Government committed to moving the nationalized industries from the public to private sector and to replacing monopoly with competition. Moreover, ESI privatization was not just a package of organizational and administrative measures, but a process of considerable political significance. As Margaret Thatcher said in her memoirs, her Government's privatization programme:

> was fundamental to improving Britain's economic performance. But for me it was also far more than that; it was one of the central means of reversing the corrosive and corrupting effects of socialism.[1]

However, the assessment of the record of the privatized ESI cannot be conducted in simple political terms, since the issues are complex. The changes that have taken place have not involved solely privatization as such. In addition to the change of ownership from a publicly owned industry to (largely) privately owned companies, there has been major restructuring. This has been more radical in England and Wales (which make up the greater part of the system), with the separation of transmission and generation, the creation of competition in generation and (progressively) in supply, and the establishment of separate regulatory regimes for monopoly distribution and transmission. In assessing the record to date, it is therefore necessary to distinguish between privatization as such, the structural changes and regulatory measures to stimulate competition, and the associated liberalization of fuel and equipment purchases.

It needs to be recognized that the actual process of privatization and restructuring of the industry represented a major administrative achievement, given the complexity of the industry and the radical nature of the changes made. Moreover, the changes were carried through in a way which met other Government objectives in the energy sector, namely to facilitate the managed rundown of the coal industry, and to sustain nuclear power (even though the original intentions to include nuclear stations in the privatization, and encourage new nuclear construction, had to be abandoned). To date, the reorganization of the industry has led to no disruption of electricity supplies, and standards of service have generally improved. By the end of the period, the balance between shareholders and consumers was being redressed in favour of the latter, prices for most consumers were falling in real terms, and environmental targets for emissions were more than achieved. The pressures for economic efficiency were increased: the upward trend in labour productivity was accelerated; the greater transparency of nuclear costs exposed the unfavourable economics of further nuclear investment; and the privatization of the ESI played a part in phasing out uneconomic coal production and creating a greater diversity of primary energy supply. The cost of capital had risen, radically altering the range of technologies for power generation which were now considered feasible (ie gas, not nuclear power).

It is important to avoid attributing solely to the 'privatization' settlement some of these benefits, which were heavily influenced by other factors. In physical terms, the most important change in the industry has been the 'dash for gas' – namely the programme of combined-cycle gas turbine (CCGT) plant, operating at high load factors, which will result in about half of the ESI's previous use of coal being replaced by gas. This has been the principal means by which the industry's primary fuel supply has been diversified and reliance on UK coal diminished. CCGTs have also emerged as the industry's preferred means of meeting environmental targets. Yet, although the 'dash for gas' was facilitated by the privatization and reorganization of the ESI, it also owed much to the emergence of a substantial surplus of gas from the UK continental shelf. Similarly, the falls in electricity prices since privatization, where they have occurred, have owed at least as much to falling coal and gas prices, as to privatization as such.

OUTSTANDING ISSUES

Against this apparently favourable record of the UK ESI since privatization (albeit aided by external factors), there need to be set a number of negative features which have given rise to as yet unresolved issues.

A disproportionate amount of the industry's economic rent has gone to shareholders in the new companies. Levels of profitability and rates of return on capital have been widely seen as excessive, particularly as most of the profits have been made from the monopoly activities of distribution and transmission, and from the coal contracts between the major generators and British Coal, which were brokered by Government. Although regulatory rules for monopoly distribution and transmission are being tightened, a satisfactory balance between returns for shareholders and prices for consumers has yet to be struck.

The original intention to include nuclear power in the initial ESI privatization meant that National Power had to be sufficiently large to accommodate the financial risks of nuclear power, and that PowerGen also had to be sufficiently large to compete with National Power. When nuclear power was withdrawn from the initial privatization at a relatively late stage, there was no time to 'unscramble' the resultant duopoly in generation in England and Wales. Although the initial dominance of National Power and PowerGen in generation has been subsequently eroded in terms of market share by the new independent power producers, effective price competition in generation has not yet been established and an appropriate mechanism for new investment, to safeguard security of supply, has not yet emerged.

There are also important items of unfinished business. The very complex task of establishing competition in electricity supply to the domestic market, which comprises about half the total demand for electricity, has still to be achieved. It is still unclear whether this can be accomplished in a way which will give benefits to the generality of domestic customers, with more 'winners' than 'losers' and without prejudice to supply security and important social obligations. The planned privatization of nuclear power stations, but excluding the older Magnox stations, has (as of May 1996) not yet been carried through. Finally, from the outset the structure of the industry did not entirely reflect the distinction between monopoly and competitive activities, and subsequent changes have led to further divergence from the theoretical model of vertical separation. Following the takeover activity in 1995/96, it is already clear that the future structure of the industry will differ greatly from that laid down at the time of privatization. Within a year of takeovers and mergers of RECs being allowed, half of the RECs ceased to have an independent existence, and further changes are likely. Vertical integration has emerged as a major continuing issue, despite the Government's decision in April 1996 not to allow the bids by National Power and PowerGen for two RECs. Moreover, the issues of integration of electricity with gas supply or other utility services and of direct or indirect foreign ownership of the ESI (particularly by US utilities), have yet to be played out.

The fact that after nearly six years there are still a number of unresolved issues should occasion no surprise. The transition from a stable public-sector framework based on monopoly to a more fluid private and liberalized structure could not have been achieved overnight. Indeed it is still unclear whether the UK ESI is in transition to a new (but different) form of stability, or whether there will be a process of continuous change in response to external factors (such as changes in the international fossil fuel market or environmental targets) or internal tensions within the industry as the 'British Electricity Experiment' evolves.

Some of these unresolved issues are more intractable than others. The shortcomings in the initial organization of generation, and the too-lenient price regulation of RECs can be remedied in time. However, there remains a fundamental uncertainty as to the overall balance of advantage. Radical change from public control to a privatized and liberalized structure involves a trade-off between greater incentives to economic efficiency and consumer choice on the one hand, and higher commercial risks on the other. Whether the UK system will provide significant price and other advantages to the generality of customers on an ongoing basis, without prejudice to long-term security of supply, remains to be seen.

Lessons

Although it is too soon to judge the balance of advantages and disadvantages of the new regime, experience during the first six years after UK electricity privatization does indicate a number of lessons of significance not only for UK policy makers, but also for those in other countries contemplating using the UK experience as a basis for reform of their electricity industries. We shall now outline the main lessons which appear to us to have emerged so far.

THE IMPORTANCE OF THE INITIAL SETTLEMENT

The more comprehensive the initial settlement in terms of organization and regulation, the greater will be the difficulty and the time required to remedy any defects which emerge. In the UK, given that ESI privatization was at the time politically very contentious, the electoral timetable dictated a comprehensive settlement (even though competition was due to be phased in over a period). The dominance of National Power and PowerGen (which has been the main obstacle to effective competition in generation) and the generous balance sheets and price controls given to the RECs (which led to excessive profits and greatly increased the problems of subsequent regulation) were in large measure the result of the Government's over-riding priority to complete the privatization process quickly and irreversibly.

GOVERNMENT INVOLVEMENT

Privatization does not end Government involvement with the ESI, as is sometimes claimed: rather the character of that involvement is changed. In the UK there has been a move away from Government involvement in the ESI's fuel-purchasing and investment policies, but with continuing oversight of the regulatory process and greater interest in environmental issues. Although the Government lost some of the policy levers which had been available when the electricity industry was in public ownership, it retained very wide residual powers under the 1989 Electricity Act. In any case, some form of political oversight is inevitable given the political risks if electricity prices are seen to be excessive or security of supply is thought to be at risk. While the 'arms-length' independence of the regulator, with minimal intervention by Government, can be sustained in the absence of such difficulties, and where there is no conflict with Government policy, in less propitious circumstances political intervention is unlikely to be avoided. Even if the Government itself would prefer not to intervene, one or more affected parties may demand that the Government steps in. The relationship between Government and regulator will always be ambiguous, particularly as there is a fundamental tension between public accountability and political interference.

The Problems of Regulation

The problems of electricity regulation are both difficult and complex, and expectations that regulation can be both light and temporary are unlikely to be fulfilled. Moreover, it is difficult to achieve a regulatory framework that is capable of providing both stability and responsiveness to changing circumstances. Above all, the structure and regulatory framework have to pass the test of public acceptability if they are to be sustainable.The scope of electricity regulation has been particularly ambitious in the UK. The regulator has been charged not only with the traditional duty of controlling monopoly, but also given a primary duty to promote and manage the evolution of competition. In the case of the monopoly transmission and distribution activities, there is a lack of objective tests with which to determine an equitable and politically acceptable share of economic benefits between consumers and investors. It is also difficult to ensure that the investment programmes for the monopoly networks are both adequate and cost-effective. These problems are increased because the companies will generally have a great deal more of the relevant information than the regulator, particularly where the monopoly activities form only part of the activities of the companies concerned.

Competition and Industry Structure

Given that the initial privatization structure can change rapidly once takeovers and mergers are allowed, regulation to promote and control competition requires from the outset a clear view of an acceptable model of competition and continuing regulation to enforce it. In the absence of countervailing policies there will be a strong tendency towards consolidation, particularly aimed at reducing the risks of power generation. Changing patterns of ownership and industry structure inevitably increase the problems and complexity of regulation, especially where foreign ownership is involved.

COMPETITION IN SUPPLY

The potential benefits of competition in supply to final customers, and how best to achieve them, need to be thought through systematically. There are dangers if the promotion of competition is seen as an end in itself or if competition is seen as self-evidently a good thing. The likely balance between costs and benefits and the ease with which competition can be introduced vary widely between different sectors of the market. Competition is easiest to introduce for large industrial and commercial users, and most difficult for the domestic market, where the transactions and settlement arrangements are likely to prove to be complex and costly, and where the elimination of cross-subsidies and erosion of social obligations raise potential political difficulties unless appropriate action is taken. Domestic-market competition is unlikely to reduce prices to the average domestic consumer unless it has the effect of creating more effective competition in generation than would otherwise have been secured.

COMPETITION IN GENERATION

The key to providing price advantages from competition to final consumers – whether domestic, commercial or industrial – lies overwhelmingly in generation by reducing generators' margins and costs within the wholesale price of electricity. Other components of electricity costs are either small, in the case of supply, or regulated, in the case of transmission and distribution. UK experience shows that open access to the network for new entrants is a necessary but not sufficient condition for effective competition in generation. It is primarily the structure of the ESI, particularly the number and plant profiles of the generators, the extent of excess generating capacity and the presence of supply competition, that will determine whether effective price competition in generation can be established. Once effective price competition is established, the commercial risks of power-station investment, and therefore the generators' cost of capital, will be markedly increased, and this could have implications for long-term supply security. The use of high discount rates sits uneasily with the need to give greater weight to the longer term on grounds of sustainability.

ESI and the Primary Energy Market

The importance of the relationship between the ESI and the primary energy market can hardly be exaggerated, as primary fuel costs are by far the most important single element in the whole chain of costs to final electricity customers, and can vary widely through time depending on market conditions. In the first six years following ESI privatization, fossil fuel markets have generally been over-supplied and low-priced leading to substantial reductions in the price of coal and gas supplied to the ESI. In turn, these would have almost certainly led to electricity price reductions comparable to those actually achieved, even if ESI privatization had not taken place. In the event, coal and gas price reductions enabled electricity price reductions to be presented as benefits flowing from privatization and the introduction of competition. The political acceptability of privatization has been greatly dependent on this fact. However, the progressive increase of competitive pressures on electricity generators and suppliers has increased risks to fuel suppliers: the main casualties are the British deep-mined coal industry and nuclear power (with renewable energy sources sustained only by special subsidy and market protection). Indeed, UK experience has shown the extreme difficulty of privatizing nuclear power in a way compatible with new investment, particularly if privatization is associated with competition and discouragement of vertical integration.

Long-Term Uncertainty

It is still not clear whether the British Electricity Experiment can deliver consumer benefit and supply security on a sustainable basis. In a fully competitive market, individual companies lose control of capacity planning and hence can no longer be held responsible for long-term security of supply. A fully competitive electricity market, replacing the former integrated framework for load forecasting and capacity planning is likely increasingly to have the characteristics of a commodity market, using financial instruments rather than long-term physical contracts. An electricity system of this kind, dominated by the financial markets, could well lead towards short-termism and instability. Such tendencies are likely to be reinforced by the nature of the UK financial markets, which are widely recognized as being preoccupied with high rates of return and short-term cash flow. It is very unclear how the new competitive system would adapt to a fundamental and adverse change in the supply conditions for primary fuel or whether its operation can be made compatible with supply security or optimal environmental policies in the longer term. The question will be given further force if the ESI becomes even more dependent on gas, since the electricity and gas industries have a similar peak-demand profile.

RELEVANCE OF UK EXPERIENCE TO OTHER COUNTRIES

Other countries seeking to use the experience of UK electricity privatization need to bear in mind that this is very much the product of UK circumstances. The UK electricity industry before privatization, despite the complexity of its organization, operated as an integrated system. Apart from Scotland and Northern Ireland, the regional dimension was unimportant. The industry was also mature, with established infrastructure, only modest demand growth and adequate capacity. UK experience would thus be of doubtful relevance for countries with strongly decentralized or regionally based electricity systems where the primary fuel resource base is quite different, or those where the priority was to create sufficient infrastructure and generating capacity to allow rapid expansion in electricity use for economic and social development.

UK experience has also been heavily influenced by UK political factors. ESI privatization was a politically contentious issue, with an associated political agenda – most notably in relation to the run-down of the coal industry. It was carried through because of the power of central Government under the British system. Political considerations precluded gradualist or hybrid solutions – for example a publicly owned monopoly infrastructure, with privately owned generation. It is unlikely that UK experience would be wholly relevant to countries with strong regional Governments, or for those wishing to create separate timetables for restructuring and privatization as such. The privatization and restructuring of the ESI in the UK is not only an Experiment, it is very much a *British* Experiment, which is still in progress.

Notes and References

Chapter 1 Introduction

1 G Yarrow (1995), 'Privatization, Restructuring and Regulatory Reform in Electricity Supply', in M Bishop, J Kay, and C Meyer (eds), *Privatization and Economic Performance*, Oxford University Press, pp 62–88. See especially section 2: 'Underlying Issues'.

Chapter 2 UK Electricity Supply under Public Ownership.

1 *Brighton and the Electric Revolution, 1882–1982* (1985) Central Electricity Generating Board, South Eastern Region, Ref 2M/1/85.
2 *Report of the Committee to Review the National Problem of the Supply of Electricity* (the Weir Committee) (1925) HMSO, London.
3 CEGB Memorandum, House of Commons Energy Committee, *The Structure, Regulation and Economic Consequences of Electricity Supply in the Private Sector*, Third Report, Session 1987–88, HMSO, London, HC 307-II, p 9.
4 Op cit, HC 307-I, p xiii.
5 For further details see the Monopolies and Mergers Commission's reports on *The South of Scotland Electricity Board*, Cmnd 9868, 1986; and on *The North of Scotland Hydro-Electric Board*, Cmnd 9628.
6 See *Electricity Prices in Northern Ireland (1995)*, House of Commons Northern Ireland Affairs Committee, Second Report, Session 1994–95, HC 395, HMSO, London, November.
7 *Report of the Committee of Inquiry into the Electricity Supply Industry* (the Herbert Report) (1956), Cmnd 9672, HMSO, London.
8 P Lesley Cook and A J Surrey (1977), *Energy Policy: Strategies for Uncertainty*, Martin Robertson, London, p 133.
9 *The Structure of the Electricity Supply Industry in England and Wales* (the Plowden Committee Report) (1976), Cmnd 6388, HMSO, London.
10 This excludes a number of small, 'off-shore' utilities on the main UK islands: Alderney Electricity Ltd, Jersey Electricity Co Ltd, the Manx Electricity Authority (Isle of Man), and the States of Guernsey Electricity Board.

11 *The SSEB* (1986), MMC, Cmnd 9868, HMSO, London, pp 111–127.
12 Op cit HC 307-I, p xiv.
13 Op cit HC 307-II, p 10, para 11.
14 *Ministry of Fuel and Power Act, 1945*, Section 1, para 1.
15 Op cit (the Plowden Report), p 9, para 2.21.
16 White Paper (1961), *The Financial and Economic Obligations of the Nationalized Industries,* Cmnd 1337, HMSO, London.
17 White Paper (1967), *The Nationalized Industries: A Review of Economic and Financial Objectives*, Cmnd 3437, HMSO, London.
18 White Paper (1978), *The Nationalized Industries*, Cmnd 7131, HMSO, London.
19 White Paper on Cash Limits (1975), Cmnd 6440.
20 A Holmes, J Chesshire and S Thomas (1987), *Power on the Market: Strategies for Privatizing the UK Electricity Industry*, Financial Times Business Information, London, pp 92–99.
21 *Power and Money: Financial Aspects of Electricity Privatisation (1988)*, Privatisation Discussion Paper No 6, Electricity Consumers' Council, London. Comparison of historic cost accounting (HCA) and current cost accounting (CCA) rates of return is difficult. However, for example, the 4.1 per cent return on CCA assets achieved by AEBs in 1986–87 was equivalent to a 20.3 per cent rate on HCA assets, similar to that achieved in private industry.
22 Select Committee on Energy, *Electricity and Gas Prices*, First Report, Session 1983–84, HC 276-I, pp xxxvi–xxxvii.
23 In fact reassessment of the provisions for nuclear fuel reprocessing, nuclear waste management and decommissioning liabilities resulted in a large and sudden increase in liabilities for both the CEGB and the SSEB. Technically these made the SSEB bankrupt and it needed special Government support. In the case of the CEGB the value of liabilities rose by over £4600 m in FY 1988/89. See *CEGB Annual Report and Accounts*, 1988/89, p 46, note 18.
24 National Board for Prices and Incomes (1968), *The Bulk Supply Tariff of the CEGB*, Report No 59, Cmnd 3575, HMSO, London; and the Price Commission (1978), *Fuel Cost Adjustment for The Supply of Electricity*, HC 133, HMSO, London.
25 See for example the Price Commission (1978), *SSEB: Price Increases in the Supply of Electricity*, HC 535, HMSO, London.
26 For example, *Report on the Generation Security Standard (1985)*, Electricity Council, London.
27 Quoted in Monopolies and Mergers Commission (1981), *The Central Electricity Generating Board*, HC 315, HMSO, London, p 40.
28 Select Committee on Energy (1981), *The Government's New Nuclear Power Programme*, First Report, Session 1980–81, HC 114, HMSO, London; the Government's Reply was published as *Nuclear Power*, Cmnd 8317, HMSO, London (1981).

29 For further details see J H Chesshire (1992), 'Why Nuclear Power Failed the Market Test in the UK', *Energy Policy*, 20, (8), August, pp 744–754.
30 The fullest exposition of the CEGB's coal purchasing policy is set out in Select Committee on Energy (1986), The Coal Industry, Memoranda Laid Before the Committee, CEGB Memorandum, HC 196-I, HMSO, London, pp 88–108.
31 A Sherry (1984), 'The Power Game: The Development of Conventional Power Stations, 1948–83', The James Clayton Lecture, *Proceedings* 198, (174), Institution of Mechanical Engineers, London, p 8.
32 *Report of the Committee of Enquiry into Delays in Commissioning CEGB Power Stations*, (the Wilson Report) (1969), Cmnd 3960, HMSO, London.
33 Select Committee on Energy (1984), *Energy Research, Development and Demonstration in the UK*, Ninth Report, Session 1983-84, CEGB Memorandum, HC 585-II, HMSO, London, p 19 and pp 31–32.
34 Op cit, HC 307-I, p xvii, para 17.

Chapter 3 Privatization of the Electricity Supply Industry

1 See J Kay, C Mayer, & D Thompson (eds) (1986) *Privatisation & Regulation – The UK Experience*, Clarendon Press, Oxford; and J Vickers & G Yarrow (1988) op cit for accounts of the UK privatization programme.
2 For a comparison of the offer price that shares were sold to the public at and the corresponding opening see P Grout (1995) 'Popular Capitalism' in M Bishop, J Kay, C Mayer (1995) *Privatisation & Economic Performance*, Oxford University Press, Oxford, p 302.
3 Figures are taken from P Grout, op cit.
4 See in M Bishop, J Kay, C Mayer 'Introduction: Privatisation in Performance' in M Bishop, J Kay, C Mayer op cit.
5 Reported in *Power Europe*, 'CEGB: Is breaking up hard to do?', 10 October 1987.
6 A Holmes, J Chesshire & S Thomas (1987) *Power on the market: Strategies for privatising the UK electricity industry*, Financial Times Business Information, London.
7 S C Littlechild (1983) *Regulation of British Telecommunications' Profitability* Department of Industry, London.
8 In M Parker & A J Surrey (1995) 'Contrasting policies for coal and nuclear power, 1979–92', *Energy Policy* 23 (9), pp 821–850, the authors chart systematic Government discrimination against British coal and in favour of nuclear power in the period of the Thatcher Governments.
9 See for example, J H Chesshire (1989) 'Was Britain's electricity privatisation

cleverly done?', paper presented to the Financial Times World Electricity Conference, 16–17 November 1989, London.

10 Several former Government Ministers in Thatcher Governments have stressed the deep animosity between the NUM and the Government and the Government's strategic objective of breaking NUM power. See for example, N Ridley (1991) *My Style of Government*, Hutchinson; P Walker (1991) *Staying Power*, Bloomsbury Publishing; C Parkinson (1992) *Right at the Centre* Weidenfeld & Nicolson; and N Lawson (1992) *The View from No 11* Bantam Press.

11 N Lawson, op cit, p 168.

12 For example, many of the nuclear power plants were operated at up to 10 per cent above their nominal capacity during the winter of 1984/85. *See Operating Experience with Nuclear Power Stations in Member States in 1985* International Atomic Energy Agency, 1986, Vienna.

13 *Privatising Electricity: The Government's Proposals for the Privatisation of the Electricity Supply Industry in England and Wales*, (1988) Cm 322, HMSO, London. The proposals for Scotland were published in *Privatisation of the Scottish Electricity Industry* (1988), Cm 327, HMSO, London.

14 A Holmes, J Chesshire & S Thomas, op cit, p 64.

15 Third Report, Session 1987–88, HC 307-1, HMSO, London, July 1988.

16 The initial CEGB case to build Hinkley Point C stated: 'the policy for diversity justifies consent for and the construction of generating plant which would be needed to meet the requirement for non-fossil-fuelled capacity. It is the CEGB's case that Hinkley Point 'C' is essential to meet that requirement.' Central Electricity Generating Board (1988) *Hinkley Point 'C' Power Station Public Inquiry, Statement of Case*, p 10, para 1.13. The CEGB sought to avoid comparing the cost of power from a PWR with that of a coal-fired station, but the inquiry inspector concluded that: 'using a 10% discount rate [the rate he concluded was appropriate for the private sector],there would be a significant advantage in favour of coal fired plant.' *The Hinkley Point Public Inquiries: A report by Michael Barnes QC* (1990), vol 3, p 856, para 29 138, HMSO, London.

17 In evidence to the Energy Select Committee, Wakeham said, 'If I was starting from scratch I would not have decided to split the CEGB fossil stations into two companies' House of Commons Select Committee on Energy, *The Cost of Nuclear Power*, fourth report, session 1989–90, HC 205-I, p x, para 6.

18 For a review of the conduct of the privatization of the RECs, see National Audit Office (1992) *The sale of the twelve regional electricity companies*, HC 10, HMSO, London.

19 Most of the grid operates at 275 kV or 400 kV. The distribution system, owned by the RECs, operates at 132 kV or less.

20 For a review of the conduct of the sale of the two large generation companies, see National Audit Office (1992) *The Sale of National Power and PowerGen*, HC 46, HMSO, London.
21 See L Hannah (1979) *Electricity before Nationalisation: A Study of the Development of Electricity Supply in Britain to 1948*, Macmillan, London.
22 For a review of the conduct of the privatization process, see National Audit Office (1992) *The Sale of Scottish Power and Hydro-Electric*, HC 113, HMSO, London.
23 *Privatisation of Northern Ireland Electricity*, (1991) Cm 1469, HMSO, London.
24 For a review of the conduct of the privatization process, see Northern Ireland Audit Office (1994) *The Privatization of Northern Ireland Electricity*, Cm 667, HMSO, London.
25 For a review of developments in the Northern Ireland electricity supply system since privatization, see Northern Ireland Affairs Committee (1995) *Electricity Prices in Northern Ireland*, CM 395-1, HMSO, London.

Chapter 4 The Development of Competition

1 In some respects, this continued previous practice whereby very large consumers were sometimes directly supplied under a negotiated contract by the CEGB.
2 Referred to in L Phillips (1995) *European Electricity Liberalisation: Lessons from the UK*, MC Securities, London, p 12.
3 Department of Trade and Industry and the Scottish Office (1995) *The Prospects for Nuclear Power in the UK*, Cm 2860, HMSO, London.
4 An alternative approach was adopted for the Norwegian Power Pool, allowing contracting which bypassed the Pool, leaving the Pool as a clearing mechanism for marginal power. Bottlenecks in the grid are overcome by a separate bidding process.
5 VOLL has risen with inflation and in 1996, was about £2.20.
6 Because of the characteristics of these plants, they are unlikely to be able to set the marginal price for more than about 20 per cent of the time.
7 See L Phillips, op cit, p 7.
8 The cost and profit data quoted in this section are drawn from L Phillips, op cit, pp 6–8.
9 See L Phillips, op cit, p 7.
10 See L Phillips, op cit, p 7.

Chapter 5 Regulation

1 OFFER (1995) *Annual Report 1994*, HC 432, 6 June, p 75; figure 38, p 76.
2 *Electricity Act 1989* (1989) Chapter 29, HMSO, p 1, para 1 (1).
3 R Trotman *Experience with Utility Regulation in Great Britain* paper presented to UN Workshop on Regulation , Buenos Aires, 27–29 September 1995, appendix 2.
4 See OFFER (1995) *Annual Report 1994* HC 432, 6 June, chapter 5.
5 *Electricity Act 1989*, chapter 29, pp 1–14.
6 *Electricity Act 1989*, chapter 29, paras 3 (1), (2), and (3), pp 2–3.
7 A Khan (1970) *The Economics of Regulation Principles and Institutions* John Wiley, New York.
8 OFFER (1994) Submission to the Nuclear Review, October, pp 9–13.
9 The *Electricity Act 1989* chapter 29, section 12, pp 9–11, specifies circumstances in which the DGES can call in the MMC.
10 M Beesley and S Littlechild (1994) 'Privatization: Principles, Problems and Priorities' reprinted in M Bishop et al (eds) *Privatisation and Economic Performance* Oxford University Press.
11 I Viehoff (1995) Evaluating RPI–X NERA Topics 17, National Economic Research Associates, London, June, pp 2–4, 8–10.
12 G MacKerron (1995) 'Regulation and the Economic Outcomes of Electricity Privatisation in England and Wales' *Revue de l'Energie* Jan/Feb, pp 77–83.
13 Secretary of State for Energy (1988) *Privatising Electricity* Cm 322, HMSO, February, paras 50–52, pp 13–14.
14 G MacKerron (as note 12).
15 House of Commons Trade and Industry Committee (1995) *Aspects of the Electricity Supply Industry* 11th Report, Session 1994–95, HC-481-1, 19 July, Table 1, p ix. The data in the rest of this paragraph are from the same source.
16 OFFER (1994) *Annual Report 1993* 352, 17 May, p 5.
17 OFFER (1993) *The Supply Price Control: Proposals* July.
18 OFFER (1994) *Annual Report 1993* 352, 17 May, pp 29–30.
19 House of Commons Trade and Industry Committee (as note 15), para 82, p xxxiv.
20 OFFER (1994) *The Distribution Price Control: Proposals* August.
21 OFFER (1994) *The Scottish Distribution and Supply Price Controls: Proposals* September.
22 OFFER (1995) *Annual Report 1994* HC 432, 6 June, p 11.
23 MMC (1995) *Scottish Hydro-Electric: A Report on a Reference under Section 12 of the Electricity Act* HMSO, 15 June.
24 OFFER (1995) *The Distribution Price Control: Revised Proposals* July.

25 House of Commons Trade and Industry Committee (as note 15) pp xxv–xviii.
26 House of Commons Trade and Industry Committee (as note 15) pp xxxvii–xxxviii.
27 Secretary of State for Energy (1988) *Privatising Electricity* Cm 322, February, para 59, p 15; and para 66, p 16.
28 OFFER (1991) *Report on the Pool Price Inquiry* December.
29 For example in OFFER (1992) *Review of Pool Prices* December.
30 OFFER (1993) *Annual Report 1992* 646, 24 May, pp 10–11.
31 OFFER (1994) *Annual Report 1993* 352, 17 May, pp 8–10.
32 OFFER (1992) *Review of Economic Purchasing* December; and OFFER (1993) *Review of Economic Purchasing: Further Statement* February.
33 OFFER (1994) *Report on Trading Outside the Pool* July.
34 Though OFFER have published some material on the issues, including *The Competitive Electricity Market from 1998* January 1995.
35 House of Commons Trade and Industry Committee (as note 15), p xiii, paras 1–8.
36 Royal Commission on Environmental Pollution (1976) *Fifth Report* Cm 6371.
37 Commission of the European Communities (1984) 'Council Directive on Combating of Air Pollution from Industrial Plants' (84/360/EEC) *Official Journal of the European Communities* no. L, 188/20, 5 July.
38 Commission of the European Communities (1988) 'Council Directive on the Limitation of Emissions of Certain Pollutants into the Air from Large Combustion Plants' (88/609/EEC) *Official Journal of the European Communities* no. L, 366/1, 7 December.
39 Department of the Environment/Scottish Office (1995) *Environment Act 1995*, Chapter 25, HMSO.
40 *ENDS Report* (1995) 248, September, pp 35–37.
41 Commission of the European Communities (as note 37), Article 4.
42 Royal Commission on Environmental Pollution *Eleventh Report*, Cm 310 HMSO.
43 Department of the Environment and the Welsh Office (1993) *Integrated Pollution Control: a Practical Guide* HMSO, pp 11–13.
44 Department of the Environment (1995) *Climate Change. The UK Programme Progress Report on Carbon Dioxide Emissions* December.
45 Department of the Environment (as note 44), figure 10, p 25.
46 Commission of the European Communities (as note 38), annex 1, p 7.
47 House of Commons Trade and Industry Committee (1993) *British Energy Policy and the Market for Coal* First Report, Session 1992–93, HC 237, 26 January, p 41, para 62.
48 House of Commons Trade and Industry Committee (as note 47), p 41, para 64.

49 UN Economic Commission for Europe (1994) *Protocol to the 1979 Convention on Long-Range Transboundary Air Pollution and Further Reductions of Sulphur Emissions.*
50 House of Commons Trade and Industry Committee (as note 47), p 42, para 66.
51 HMIP (1996) *ESI Review by HMIP Background Brief* March 26.
52 Calculations by Steve Sorrell of the Environment Programme, SPRV.

Chapter 6 Effects on Demands for Fossil Fuels

1 N Lawson (1992) *The View from Number 11* London: Bantham Press p 161.
2 C Parkinson (1992) *Right at the Centre* London: Weidenfield and Nicolson p 280.
3 Ibid p 281.
4 M Thatcher (1993) *The Downing Street Years* London: Harper Collins p 340.
5 Ibid p 378.
6 Parkinson, op cit, p 281.

Chapter 7 Nuclear Power under Review

1 See for instance A Holmes et al. *Power on the Market* Financial Times Business Information, May 1987, and A Henney *Privatise Power: Restructuring the Electricity Supply Industry – the Full Version*, May 1987.
2 Unlike most other chapters in this book, attention is not confined here to England and Wales, and Scotland is given more extensive treatment. This is mostly because the proposed new structures for nuclear power combine ownership across the English/Scottish border, unlike the situation for any other part of the England and Wales industry.
3 See J Chesshire 'Why Nuclear Power Failed the Market Test in the UK' *Energy Policy* August 1992, p 749.
4 Department of Trade and Industry and the Scottish Office *The Prospects for Nuclear Power in the UK* Cm 2860, HMSO, May 1995.
5 R Williams *The Nuclear Power Decisions* Croom Helm, 1980, provides a good history of civilian nuclear power in Britain up to 1979.
6 The AGR decision of 1965 is detailed in H Rush, G MacKerron and J Surrey 'The Advanced Gas-Cooled Reactor' *Energy Policy* 5 (2), June 1977.
7 See R Williams, op cit, especially pp 257–259.
8 *The New Nuclear Power Programme* Statement in the House of Commons by the Secretary of State for Energy, 18 December 1979.
9 Department of Energy *Sizewell B Public Inquiry* Report by Sir Frank Layfield,

HMSO, 1987.

10 See G MacKerron *The Capital Costs of Sizewell C* COLA Submission to the Government's Nuclear Review, Volume 3, September 1994, Table 2.1, p 3.

11 See S Goddard (then a Board member of Nuclear Electric) 'Future Programmes' in Institute of Energy Seminar Series. *Where are we now on Nuclear Power?*, especially graph on p 69, in which it is shown that the full generating costs of Sizewell B will be 5p/kWh (using an 8% discount rate), compared to average Pool prices of the order of 2.5p/kWh in the 1994 to 1996 period.

12 The first (and worst) example was contained in an appendix to the CEGB's 1979/80 *Annual Report and Accounts.*

13 See G MacKerron *Nuclear Power and the Economic Interests of Consumers* Electricity Consumers' Council Research Report no 6, 1982.

14 CEGB *Analysis of Generation Costs* 1983 (referring to the financial year 1981/82), and *Analysis of Generation Costs 1983/84* Update 1985 These were collectively known as the 'Grey Books'.

15 See Appendices to the Minutes of Evidence, Appendix 2, 'Memorandum Submitted by BNFL on the Effect of BNFL Fixed Prices on Generating Costs and Nuclear Provisions' in House of Commons Energy Committee *The Cost of Nuclear Power* 4th Report Session 1989–90, Volume II, HC-205-II, 27 June 1990, Table on p 139; and para 5, p 136.

16 Derived from figures presented in A Gregory 'Decommissioning of Nuclear Power Stations' paper presented to *International Seminar on Decommissioning of Nuclear Facilities*, London, 6/7 July 1988.

17 See note 1.

18 See for instance F Jenkin *Proof of Evidence on the Need for Hinkley Point 'C' to Help Meet Capacity Requirement and the Non-Fossil-Fuel Proportion Economically* CEGB 4, Hinkley Point 'C' Power Station Public Inquiry, September 1988.

19 These figures finally emerged in evidence to the House of Commons Trade and Industry Select Committee in late 1992. See House of Commons Trade and Industry Committee *British Energy Policy and the Market for Coal* First Report Session 1992–93, HC 237, 26 January 1993, Table 18 and para 120, p 61.

20 See G MacKerron *Implications of the Attempted Privatisation of Nuclear Power in Britain for Nuclear Costs* Paper prepared for Coalition of Environmental Groups and presented to Ontario Environmental Assessment Board, February 1993, especially pp 9–12.

21 Lord Marshall, in a celebrated farewell speech to the industry *The Future for Nuclear Power* BNES Annual Lecture, Royal Lancaster Hotel, 30 November 1989, outlined the causes of this escalation, See especially pp 21–24 of this speech.

22 See House of Commons Energy Committee *British Nuclear Fuels plc: Report and Accounts 1987–88*, Third Report, Session 1988–89, HC 50, HMSO, 5

April 1989.

23 See reference at note 15 above, para 9, p 135.
24 On the rise to a 10% discount rate, and the 'private sector prudent' assumptions, see Lord Marshall (reference in note 21 above), especially p 32.
25 See Lord Marshall (reference in note 21 above), Table on p 32, and G MacKerron (reference in note 20), Table 4, p 20.
26 'Memorandum Submitted by the Department of Energy' in House of Commons Energy Committee (reference in note 15), Volume II, p 23, paras 13–17.
27 The discussion in this paragraph, including all the price levels, is based on House of Commons (reference in note 15), Volume I, pp xxii–xxiv.
28 House of Commons (reference in note 15), Volume I, p x, para 6.
29 Letter from the Secretary of State for Energy to Mr Michael Clark, Chairman of the House of Commons Energy Committee, June 1990.
30 Nuclear Electric *Report and Accounts 1994/95*, chart on p 3.
31 See House of Commons Trade and Industry Committee (reference in note 19), p 59, para 111.
32 Ernst and Young *Review of Magnox Avoidable and Unavoidable Costs* Coal Review, Department of Trade and Industry, 5 February 1993.
33 See Department of Trade and Industry (reference in note 4), para 7.12, p 50.
34 Information in this paragraph is derived from Scottish Nuclear *Securing Our Energy Future* Submission from Scottish Nuclear Limited to the Government's Nuclear Review, July 1994, pp 4–6.
35 Scottish Nuclear *Annual Report and Accounts 1994/95* Five Year Financial Summary.
36 Scottish Nuclear (reference in note 34), para 11, p 4.
37 Information in this paragraph is largely from House of Commons Trade and Industry Committee (reference in note 19), pp 61–65.
38 M Heseltine, President of the Board of Trade, Hansard 19 October 1992.
39 House of Commons Trade and Industry Committee (reference as in note 19), para 121, p 62.
40 Nuclear Electric *Report and Accounts 1994/95* Profit and Loss Account, p 47.
41 These sums do not exactly correspond to the sums shown in Table 7 1, mainly because Table 7 2 is in current not constant money values.
42 Nuclear Electric *Report and Accounts 1994/95* p 31.
43 Nuclear Electric *Report and Accounts 1994/95* p 31,.
44 OFFER *Submission to the Nuclear Review* October 1994, pp 19–21.
45 Nuclear Electric *Report and Accounts 1994/95* Chart on p 3, referring to 1989/90 operating costs.
46 This figure refers to 1990/1991. See Scottish Nuclear *Annual Report and Accounts 1994/95* Five Year Financial Summary, p 58.
47 McGraw-Hill *Nucleonics Week* February 8 1996, pp 5-6.

48 Nuclear Electric *Station Performance Survey 1994–95* Summary of Operating Experience in Nuclear Power Plants, shows that seven out of the ten Nuclear Electric AGR reactors (two per station) have unplanned 'automatic trips' (shutdowns) well above the world median values, with none below those values.
49 Data in this paragraph are from Nuclear Electric's and Scottish Nuclear's *Annual Reports and Accounts* 1990/91 and 1994/95.
50 See G MacKerron and M Sadnicki *UK Nuclear Privatisation and Public Sector Liabilities* STEEP Special Report no 4, Science Policy Research Unit, University of Sussex, November 1995, pp 11–14.
51 Financial Times Energy Publishing Power In Europe, 7 April 1995, 196, p 3.
52 Scottish Nuclear (reference in note 34), p 14.
53 Data in this paragraph are from Nuclear Electric, various *Annual Reports and Accounts*, and from Scottish Nuclear, various *Annual Reports and Accounts*, especially 1994/95.
54 House of Commons Trade and Industry Committee (reference as in note 19), para 131, p 65.
55 Department of Trade and Industry (reference as in note 4).
56 Department of the Environment *Review of Radioactive Waste Management Policy Final Conclusions*, Cm 2919, July 1995.
57 Department of Trade and Industry (reference as in note 4), pp 3–6.
58 Department of Trade and Industry (reference as in note 4), chapter 4, pp 17–25.
59 Department of Trade and Industry (reference as in note 4), chapter 5, pp 27–30.
60 Department of Trade and Industry (reference as in note 4), chapter 5, pp 31–36.
61 Department of Trade and Industry (reference as in note 4), paras 2.10 and 2.11, p 4.
62 OFFER *Submission to the Nuclear Review* October 1994, pp 9–13.
63 Department of Trade and Industry (note as in reference 4) para 9.20, p 64.
65 OFFER, Letter to House of Commons Trade and Industry Committee Enquiry into Nuclear Privatisation, 9 January 1996.
64 House of Commons Trade and Industry Committee *Nuclear Privatisation* Minutes of Evidence, 10 January 1996, 43-iii, question 590, p 123.
66 The DTI press release accompanying the Nuclear Power White Paper in (9 May 1995) contained a table in an annex which showed, as a residual, the sum of £2.6 bn as the sale proceeds from the AGRs and PWR.
67 G MacKerron and M Sadnicki (reference as in note 50) Table 11, p 44.
68 G MacKerron and M Sadnicki (reference as in note 50) pp 59–66.
69 See British Energy response to written questions from the House of Commons Trade and Industry Committee, Questions of January 11 1996, Question 4.

70 G MacKerron and M Sadnicki (reference as in note 50) p 75.
71 DTI Press Release Annex on Magnox Liabilities, May 9 1996.
72 DTI Press Release Annex on Magnox Liabilities, May 9 1996.

Chapter 8 Renewable Generation – a Success Story?

1 DTI (1994) *Wardle Makes Third Renewable Energy Order*, Press Release, DTI, 20 December.
2 National Audit Office (NAO) (1994) *The Renewable Energy Research Development and Demonstration Programme*, HMSO, p 7.
3 House of Commons Energy Select Committee (1992) *Renewable Energy*, vol 2, p 15: 5.
4 For more detail see C Mitchell (1995) 'The Renewable NFFO – A Review' *Energy Policy* December, 23, (12), pp 1077–1091.
5 For a description of the nuts and bolts workings of the NFFO see C Mitchell (1996) 'The UK's Non-Fossil Fuel Obligation – results and lessons', *IEPE Journal*, Bocconi University, Italy.
6 MW DNC = Mega Watt Declared Net Capacity. The DNC of a non-fossil fuel generating station 'is the highest generation of electricity which can be maintained indefinitely without causing damage to the plant less so much of that capacity as is consumed by the plant'. For value of each renewable technology see DTI (1995) *Renewable Energy Bulletin* No. 6, Annex B.
7 House of Commons Select Committee on Welsh Affairs (1994) *Wind Energy*, Vol 1, HMSO, 15 July.
8 *Windirections*, XIII, (4), 1994: 13.
9 House of Commons Environment Select Committee (1994) *Recycling*, p xxxi, para 54, HMSO, 6 July.
10 Friends of the Earth (1994) 'Small-scale Wind projects and New Wood for Fuel Plants Receive Boost', Press Release, 20 December.
11 DTI (1996) 'Renewed Interest in Renewables', Press Release, 7 March.
12 For a more detailed discussion see C Mitchell (1995) *Renewable Energy in the UK – Financing Options for the Future*, available from the Council for the Protection of Rural England, London, UK.

Chapter 9 The Winners and Losers So Far

1 Secretary of State for Energy *Privatising Electricity* The Government's Proposals for the Privatisation of the Electricity Supply Industry in England and Wales, Cm 322, February 1988, para 66, p 16.

2 Reference as in note 1, para 66 p 16.
3 Reference as in note 1, para 66, p 16.
4 OFFER *Report on Customer Services 1994/95* 1995.
5 Derived from data in National Grid Company *Seven Year Statements* March 1991 and March 1995.
6 Data on number and duration of supply interruptions are from OFFER *Annual Report 1994* HC 432, 6 June 1995, figures 20 and 21, p 52.
7 Data in this paragraph are from OFFER *Annual Report 1994* HC 432, 6 June 1995, figures 18 and 19, pp 50 and 51 respectively.
8 Information in this section is based on OFFER *Report on Customer Services 1994/95*, Chapter 2 (for Guaranteed Standards) and Chapter 3 (for Overall Standards).
9 OFFER *Annual Report 1994* HC 432, 6 June 1995, p 64, and OFFER *Annual Report 1992* 646, 24 May 1993, p 61.
10 OFFER *Annual Report 1994* HC 432, 6 June 1995, p 70; and figure 35, p 71.
11 OFFER *Annual Report 1994/95* HC 432, 6 June 1995, p 43.
12 Factual information in this paragraph is largely derived from OFFER *Energy Efficiency: Standards of Performance* 1994, especially pp i and ii.
13 D Heald 'A financial autopsy on the CEGB' *Energy Policy* 17, (4), August 1989, especially pp 348/349.
14 *The Guardian* 9 December 1995.
15 Data here from OFFER *Annual Report 1994* HC 432, 6 June 1995, figure 7, p 34.
16 Data are from Department of Trade and Industry *Energy Trends* table 28, (monthly).
17 The correspondence is not exact because in the regulatory rules, consumers have qualified for choice of supplier according to their maximum demand, whereas the DTI data reflect total annual consumption.
18 Figures calculated from daily quotations in the *Financial Times.*
19 *Financial Times* 22 December 1994.
20 These are the accounts which must be submitted annually to OFFER. They are designed to distinguish between the distribution (monopoly) and supply (partially competitive) businesses of the RECs, and help the regulator to try and ensure that there are no cross-subsidies between the two types of business.
21 These fluctuations are generally concerned with the complex issue of accounting for the very large liabilities attaching to the nuclear stations.
22 David Newbery 'The Restructuring of UK Energy Industries: What have we Learned?' Keynote address to BIEE/Warwick University Conference The UK Energy Experience: a Model or a Warning? Warwick University, 11–12 December, p 16 (and in G MacKerron and P Pearson (eds) *The UK Energy Experience: a Model or a Warning?* Imperial College Press, 1996).

23 House of Commons Trade and Industry Committee *British Energy Policy and the Market for Coal* HC 237, First Report Session 1992–93, 26 January 1993, especially pp 19–35.
24 At the end of 1992, generators' coal stocks stood at 32.2 m tonnes, and fell to 14.1 m tonnes by end-1994 (See DTI *Energy Trends* January 1996, table 6) At £34.50/tonne this reduction in stocks represented a value of around £600 m.
25 Financial Times Energy Information *Coal UK* gives monthly figures for steam coal imports: for 1993/94 the total was 6.24 m tonnes.
26 This represents a typical figure for coal import prices at the time, around a fairly wide variation.
27 House of Commons Trade and Industry Committee *Aspects of the Electricity Supply System* 11th Report, Session 1994–95, HC 481-1 19 July 1995, para 33, p xviii.
28 Reference as in note 27, para 33, p xviii.
29 *Annual Report and Accounts* various RECs.
30 *Annual Report and Accounts* CEGB 1988/89: National Power 1994/5; and PowerGen 1994/95.
31 *The Observer* 14 October 1995.
32 *The Guardian* 9 December 1995.
33 Data from Electricity Council *Annual Report and Accounts 1987/88.*
34 Data from House of Commons Committee of Public Accounts *The Sale of the Twelve Regional Electricity Companies* Session 1992–93, HC 101, p v (summary of conclusions).
35 See J Roberts et al *Privatising Electricity: the Politics of Power* Belhaven 1991, p 82, and *The Guardian* 6 February 1995.
36 The extent to which higher proceeds could have been obtained from the sale is not covered here, although the rapid growth in share values following privatization certainly suggests under-pricing on a significant scale.
37 Nuclear Electric *Report and Accounts 1994/95*, p 31.
38 British Coal Corporation *Annual Report and Accounts 1994/95* Summary of Statistics 1947–1994/95, p 36.
39 Department of Trade and Industry *Digest of UK Energy Statistics 1995* HMSO 1995, Table C10, p 187.
40 Reference as in note 34, p 182.
41 See I Boira-Segarra ' Industrial organisation and environmental management in the electricity supply industry of England and Wales' paper presented at BIEE/ Warwick Conference on *The UK Energy Experience: a Model or a Warning?* held at Warwick University, 11–12 December 1995, especially pp 6–9, (and forthcoming in G MacKerron and P Pearson *The UK Energy Experience: a Model or a Warning?* Imperial College Press, 1996).
42 National Power *Report and Accounts 1994/95* Five Year Financial and Statistical Summary, p 59.

Chapter 10 Competition: the Continuing Issues

1 House of Commons Trade and Industry Committee (1995) *Aspects of the Electricity Supply Industry*, 11th Report, Session 1994–95, HC-481-1, 19 July, HMSO, London, Para 26.
2 DTI Press Notice, P/96/830, November 1995.
3 *The Economist*, 13 April 1996.
4 DTI Press Notice, P/96/373, April 1996.
5 *The Financial Times*, 3 May 1996.

Chapter 11 Unresolved Issues of Economic Regulation

1 D Corry, D Souter and M Waterson (1994) *Regulating Our Utilities*, Institute for Public Policy Research, London.
2 House of Commons Trade and Industry Committee (1995) *Government Observations on the Eleventh Report from the Trade and Industry Committee (session 1994–95) on 'Aspects of the Electricity Supply Industry'*, HC774 , HMSO, London, 18 October, p viii, para 24.
3 S Beavis and L Elliott (1996) 'Watchdogs get more bite', *The Guardian*, London, January 25.
4 House of Commons Trade and Industry Committee (1995) *Government Observations on 'Aspects of the Electricity Supply Industry'*, HC774 , HMSO, London, 18 October, p viii, para 25.
5 M Fulwood (1995) *Towards a Regulatory Contract*, Paper presented at a conference on 'Regulation and the Regulators', Park Lane Hotel, November 2–3. (The author was with British Gas.)
6 House of Commons Trade and Industry Committee (1995) *Aspects of the Electricity Supply Industry*, HC 481-1, HMSO, London, 19 July, p xxxvi, para 86.
7 'Gold-plating' is where utility companies incur far higher investment expenditures than necessary in order to increase the value of the asset base and thus increase their total profits while achieving only the rate of return allowed by the regulator. To counter this, US regulators often decide whether to allow all or only part of investment expenditures incurred to enter the asset base which determines prices.
8 M Parker and J Surrey (1994) *UK Gas Policy: Regulated Monopoly or Managed Competition?* SPRU Special Report, Science Policy Research Unit, University of Sussex.

Chapter 12 Strategic Government and Corporate Issues

1 National Coal Board (1974) *Plan for Coal* London, NCB.
2 Orimulsion is an emulsion of water and very heavy oil produced in Venezuela.
3 For a discussion of the heavy electrical industry see S Thomas & F McGowan (1990) *The World Market for Heavy Electrical Equipment*, Nuclear Engineering International Special Publications, Sutton, UK.
4 See B Epstein & R S Newfarmer (1980) *The Continuing Cartel: Report on the International Electrical Association*, Report to the Subcommittee on Oversight and Investigations of the Committee on Interstate and Foreign Commerce, United States House of Representatives, US Government Printing Office.

Chapter 13 General Conclusions and Lessons

1 M Thatcher (1993) *The Downing Street Years*, Harper Collins, p 676.

Index

18,000
x 6

108000
x .4

432000
x 30

$1296000 mill
x .8

1036800